Forensic Geotechnical and Foundation Engineering

Forensic Geotechnical and Foundation Engineering

Robert W. Day

McGraw-Hill

New York   San Francisco   Washington, D.C.   Auckland   Bogotá
Caracas   Lisbon   London   Madrid   Mexico City   Milan
Montreal   New Delhi   San Juan   Singapore
Sydney   Tokyo   Toronto

Library of Congress Cataloging-in-Publication Data

Day, Robert W.
 Forensic geotechnical and foundation engineering / Robert W. Day.
  p.   cm.
 Includes bibliographical references.
 ISBN 0-07-016444-4
 1. Evidence, Expert—United States. 2. Forensic engineering
—United States. I. Title.
 KF8968.25.D39   1998
 347.73'67—dc21   98-18043
                                                      CIP

McGraw-Hill

A Division of The McGraw·Hill Companies

1 2 3 4 5 6 7 8 9 0   DOC/DOC   9 0 3 2 1 0 9 8

ISBN 0-07-016444-4

*The sponsoring editor for this book was Larry Hager, the editing super-
visor was Peggy Lamb, and the production supervisor was Tina
Cameron. It was set in Century Schoolbook by Dina John of McGraw-
Hill's Professional Book Group composition unit.*

Printed and bound by R. R. Donnelley & Sons Company.

 *This book is printed on recycled, acid-free paper containing a
minimum of 50% recycled, de-inked fiber.*

McGraw-Hill books are available at special quantity discounts to use
as premiums and sales promotions, or for use in corporate training programs. For more information, please write to the Director of Special
Sales, McGraw-Hill, 11 West 19th Street, New York, NY 10011. Or contact your local bookstore.

*Dedicated with love to
my wife Deborah
and my parents*

Contents

Part 2 Forensic Geotechnical and Foundation Investigations

Chapter 4. Settlement of Structures   65

Chapter 5. Expansive Soil   103

Preface

For the purpose of this book, forensic engineering is broadly defined as the investigation of a damaged or deteriorated structure. Forensic engineering is different from design, because the forensic engineer must use observation, testing, and deduction to figure out the cause of the damage or deterioration. In many cases, the forensic engineer is hired as an expert witness and may need to present the results of the investigation in a court of law.

The legal aspects described in the book are those prevailing or coming into current use in the United States. Many of the tools used to perform a forensic investigation can be used worldwide. The book covers the most common types of geotechnical and foundation projects involving the forensic engineer, and some topics are covered only briefly or omitted. For example, environmental problems (such as the leakage of municipal landfills and failure of septic systems) are not discussed because this is such a broad topic that it would require extensive treatment.

Attorneys state that in giving testimony to a jury, "Keep it simple." It is likely that the forensic expert who proves a point by using a couple of photographs will prevail over the testimony of a forensic expert who has reams of calculations. The most brilliant engineer is not necessarily the best forensic expert: rather, the pleasant individual who communicates the facts in simple terms to the jury will succeed. In keeping with this philosophy, the book has "kept it simple." Charts and photographs are prominent in this book, with equations and calculations kept to a minimum.

The book is divided into four separate parts. Part 1 (Chaps. 2 and 3) provides the basic procedures that can be applied to geotechnical and foundation engineering forensic investigations. Part 2 (Chaps. 4 to 8) deals with specific problems typically encountered by the forensic engineer, such as settlement, expansive soil, slope movement, and moisture intrusion problems. Part 3 (Chap. 9) provides a discussion of the forensic engineer's role in the development of repairs, and Part 4 (Chap. 10) is a concluding chapter with a presentation of strategies for avoiding civil liability. Throughout the book, case studies are presented to illustrate the diverse nature of geotechnical and foundation problems that can be encountered by the forensic engineer.

The book presents both the practical and legal aspects of forensic engineering. The topics should be of interest to both forensic and design engineers, especially Chap. 10, which provides strategies for avoiding civil liability.

Robert W. Day

Acknowledgments

I am grateful for the contributions of many people who helped to make this book possible. Several practicing engineers reviewed various portions of the text and provided valuable assistance during its development. In particular, I am indebted to Robert Brown, Edward Marsh, and Rick Walsh, who reviewed the book. Thanks also to Dennis Poland, Ralph Jeffery, and Todd Page for their help with the geologic aspects of the book; Miriam Radsliff and Danielle Owen for production assistance; and Rick Dorrah for drafting of figures for the book.

Several attorneys contributed to the development of this book. I appreciate the efforts of Billie Jaroszek, who provided the original copy of the case management order (App. B), John R. Sorensen who reviewed Sec. 1.4 and provided important details on the certificate of merit, and Michael M. Angello for his input concerning the statute of limitations.

I would also like to thank Professor Timothy Stark at the University of Illinois and Scott Thoeny for their aid with the case study in Sec. 6.6, and Professor Charles Ladd at Massachusetts Institute of Technology for his involvement with the case study in Section 7.8.1. Thanks also to Kean Tan, who provided Figs. 7.36 to 7.38, and Jim Meyer for Fig. 8.27.

Special thanks to Professor Kenneth Carper at Washington State University and Gregory Axten, president of American Geotechnical, who provided valuable support during the review and preparation of the book.

Tables and figures taken from other sources are acknowledged where they occur in the text. Finally, I wish to thank Larry Hager, Michelle Martineau, and others on the McGraw-Hill editorial staff, who made this book possible and refined my rough draft into this finished product.

1

Introduction

1.1 Definition of a Forensic Engineer

One of the earliest recorded legal codes dealing with construction was that of Hammurabi, the great Babylonian King (FitzSimons, 1986). His legal code of construction was quite simple:

> If a contractor builds a house and it collapses killing its owner, the contractor will be killed. If the son of the owner is killed, then so will be the son of the contractor.

Another example is the Napoleonic code of 1804. In this code, if a structure had a loss of serviceability within 10 years of its completion, due to poor workmanship or foundation failure, then the builder would be sent to prison. Although the 10-year statute of limitations for developers and design professionals has survived, the laws of automatic death or imprisonment have been abolished, fortunately for civil engineers.

In 1879, the Firth of Tay railroad bridge collapsed, killing over 200 passengers on a train. One theory for the failure was that God disapproved of traveling on Sunday, and thus made an example of the collapse. But forensic investigators developed alternative theories such as material deficiencies with the wrought and cast-iron trusses, and the possibility of wind loads contributing to the failure. The engineer, the newly knighted Sir Thomas Bouch, was relieved of his duties and died shortly thereafter (FitzSimons, 1986). Today, an act of God defense will not work because most failures are foreseeable.

From the simple code of Hammurabi to the act of God defense, modern legal codes have evolved and are more complex. FitzSimons (1986) describes the 1782 case of Folkes versus Chadd, a milestone in the development of modern civil law and the use of a forensic engineer.

Folkes was a landowner who built a dike around his low-lying land. Chadd was the trustee of a nearby harbor who said that the dike had eroded and silted up the harbor. Chadd obtained an order to have the dike removed and Folkes sued. In the course of the trial, the well-known engineer John Smeaton testified that, in his opinion, the dike was not responsible for the silting up of the harbor. The court ruled that Smeaton's opinion could be considered, and stated:

> Of this [matter], such men as Mr. Smeaton alone can judge. Therefore we are of the opinion that his judgment, formed on facts, was very proper evidence.

From then on, forensic engineers could be used to judge the facts and provide opinions in a court of law.

The standard job of a forensic engineer is to investigate the damage, deterioration, or collapse of a structure, determine the cause of the problem, and in many cases, develop repair recommendations. The forensic engineer may also have to determine who is responsible for the damaged or deteriorated structure. Ultimately, the forensic engineer may have to testify under oath in a court of law as to the findings of the forensic investigation.

The ASCE *Guidelines for Failure Investigation* (Greenspan et al., 1989) provide a summary of the qualifications of the forensic engineer. As indicated in this publication, there are several general qualities that a forensic engineer must have. The first and foremost is that the forensic engineer must be an expert in his or her field and have a thorough knowledge of the subject under investigation. This expert knowledge may have been acquired by advanced education and by years of practice. If the subject under investigation is not within the area of expertise of the forensic engineer, then the assignment must not be accepted.

Another important quality is that the forensic engineer must be impartial as to the cause of the problem and who is responsible for the damaged or deteriorated structure. In terms of impartiality, the forensic engineer must avoid the appearance of conflict of interest, bias, or advocacy (Carper, 1989). As indicated in the ASCE *Guidelines for Failure Investigation* (Greenspan et al., 1989), the forensic engineer must arrive at the final conclusions on the basis of sound engineering fundamentals and the evidence that has been developed during the investigation.

In App. A, the *Recommended Practices for Design Professionals Engaged as Experts in the Resolution of Construction Industry*

Disputes (ASFE, 1993) has been reproduced. This document provides valuable commentary on the obligations and responsibilities of a forensic engineer.

1.2 Types of Damage

Visible damage to buildings due to ground movements has traditionally been divided into three general categories (Skempton and MacDonald, 1956; Bromhead, 1984; Boscardin and Cording, 1989; Feld and Carper, 1997):

- *Architectural damage.* This type of damage affects the appearance of the building and is usually related to minor cracks in the walls, floors, and finishes. Cracks in plaster walls greater than 0.5 mm (0.02 in.) wide and cracks in masonry walls greater than 1 mm (0.04 in.) wide are considered to be typical threshold values that would be noticed by the building occupants (Burland et al., 1977).

- *Functional (or serviceability) damage.* This type of damage affects the use of the building. Examples include jammed doors and windows, extensively cracked and falling plaster, and the tilting of walls and floors. Ground movements may cause cracking that leads to premature deterioration of materials or leaking roofs and facades.

- *Structural damage.* This type of damage affects the stability of the buildings. Examples include cracking or distortions to support members such as beams, columns, or load-bearing walls. This category would also include complete collapse of the structure.

In addition to visible damage to structures, there could be other conditions such as hidden (or latent) damage and monetary (ancillary) losses. A latent condition refers to a hidden weakness in the structure. There may be no visible evidence of damage. A latent condition could be discovered by reviewing the design calculations, by testing substandard construction materials, or by inspecting or testing a defective structural member. Latent damage is a hidden problem, with failure waiting to happen (Greenspan et al., 1989).

An ancillary condition does not involve actual damage to the structure itself, but rather refers to damage in the form of monetary losses. For example, the owner or contractor may initiate a lawsuit because of cost overruns. Another example is the failure to complete the project on time, which could result in a lawsuit because of lost revenues.

In summary, there are five general categories of damage. Three categories (architectural, functional, and structural) refer to visual damage of the structure. The fourth category is latent damage, which is hidden damage. The fifth category refers to monetary losses (ancillary damages).

1.3 Typical Clients

Typical clients who hire engineers to perform forensic investigations are

- *Insurance companies.*   When an owner reports property damage to the insurance company, they may be obligated to investigate that damage. For example, many owners had earthquake insurance and sustained damage to their buildings during the California Northridge earthquake (magnitude 6.7, January 17, 1994). Once the claim is processed, the forensic engineer may be hired by the insurance company to determine the cause of damage (earthquake versus preexisting damage) and develop repair recommendations.

- *Independent adjusters.*   The filing of an insurance claim can be a complicated process, and some owners prefer to hire independent adjusters to assist them with the claim. The forensic engineer may then be hired by the independent adjusters to investigate the problem independently of any investigation by the insurance company.

- *Property owners.*   In many cases, the damage to a structure may not be covered by insurance and the property owner will hire the forensic engineer to assist with the rehabilitation or repair of the structure. This category could include individuals, corporations, financial institutions, real estate companies, homeowners associations (HOAs), or other types of owners in the private sector.

- *Parties involved in a lawsuit.*   The forensic engineer could be hired by the plaintiffs, defendants, cross-defendants, their attorneys, or other parties involved in a lawsuit. Although most lawsuits settle prior to trial, the forensic engineer should be prepared to present the findings of the investigation in a court of law.

- *Developers.*   It is well known that litigation is an expensive and time-consuming process. In some cases, developers will attempt to fix the problems to avoid a lawsuit. The developers will hire forensic engineers to develop repair recommendations. Miller (1993) indicates that this is a common approach when the problem is small and well defined or limited to one or two problems.

- *Governmental agencies.*   The forensic engineer could be hired by numerous state or national agencies. For example, the forensic engineer may be hired to investigate failed or degraded publicly owned facilities or asked to be involved in condemnation proceed-ings. The forensic engineer may also be hired or work for agencies that investigate failures, such as the Occupational Safety and Health Administration (OSHA) or the National Transportation Safety Board.

■ *Owners of historic structures.* There are many historic structures that are owned by private individuals, preservation societies, or governmental agencies. The repair or maintenance of a historic structure presents unique challenges to the forensic engineer. Section 7.7 presents a case study of the repair of a historic adobe structure.

1.4   Legal Process

The forensic engineer should understand the basic definitions and concepts of the legal process. The following sections provide a brief review of the legal process. A more in-depth discussion of the legal process is presented in *Expert: A Guide to Forensic Engineering and Service as an Expert Witness* (ASFE, undated).

1.4.1   Civil litigation

The civil litigation starts with the filing of a lawsuit by the plaintiff. The defendant is the individual or company that is being sued by the plaintiff. Near the start of the civil litigation, other parties who were involved with the project may be sued by the defendant, and they are known as cross-defendants. For example, a homeowner (plaintiff) may sue the developer (defendant) because of foundation damage. The developer may then sue the geotechnical engineer (cross-defendant).

The legal papers that initiate a lawsuit must be filed in the court of law that has jurisdiction. The jurisdiction could be based on such items as the location of the damaged structure, where the defendant resides, or the place where the contract was executed. The procedures that govern the civil litigation depend on the particular court of law and may differ from jurisdiction to jurisdiction.

The judge that is assigned the case has the ultimate authority on all aspects of the civil litigation. The plaintiff has the burden of proof in the lawsuit and must sustain the burden by a "preponderance of evidence." This means that more than 50 percent of the evidence must be in the plaintiff's favor.

Because of the expense and uncertainty of trial, most civil litigation settles prior to trial. Miller (1993) estimates that 95 percent of all construction defect cases settle before the date of trial.

1.4.2 Important legal terms

The following are some important legal terms:

Attorney-client privilege and attorney work product. The communications and written documents between the attorney and the client are privileged, which means that they can not be used as evidence.

The attorney may send the forensic engineer specific documents that are labeled *attorney work product*. In general, these documents are also protected and can not be admitted as evidence. An exception is when the forensic engineer is designated as an expert witness, in which case such documents may be discoverable and admitted as evidence.

Statute of limitations. The statutes of limitations vary from state to state. A statute of limitation requires that a person (plaintiff) file a lawsuit within a certain period of time after the triggering date, or the lawsuit will be barred. For example, an apartment house collapses, resulting in injuries to a tenant. The owner of the apartment house has a potential claim against the design professional for property damage to the apartment house. The tenant has a potential claim against the design professional for personal injuries. In this example, there are two statutes of limitations, the first based on the date of completion of the work, and the second based on the date of damage. As this example shows, the two statutes of limitations depend on the type of injury (property damage versus personal injury), date of discovery, and nature of the defect.

Standard of care and negligence. A contemporary definition of the standard of care is "that level of skill and competence ordinarily and contemporaneously demonstrated by professionals of the same discipline, practicing in the same locale, and faced with the same or similar facts and circumstances" (ASFE, 1993). The theory of the contemporary standard of care is based on the idea that the professional engineer's work must equal that of an average prudent engineer, who is aware of the state of knowledge at the time of design. The law allows the forensic engineer to determine the standard of care and judge the actions of the design engineer. In a lawsuit, a design engineer can be held negligent if he or she violated the standard of care which resulted in damage or injury. The procedure to determine the standard of care is listed in App. A, item 7.

Strict liability. The legal concept known as *strict liability* is also generally referred to as *liability without fault*. This concept of tort law was initially developed for persons who engaged in dangerous activities. For example, those individuals engaged in the ultrahazardous activity of selling explosives or blasting rock were held liable for losses without regard to whether or not they were negligent. Other individuals held to the legal standard of strict liability included the keepers of dangerous animals or persons who sold dangerous products (Sweet, 1970).

The modern development of strict liability dealt mainly with manufactured goods. The development of manufacturer's liability varies from state to state, but in general, strict liability for a manufacturer means that they are held liable for injuries caused by their defective products, and the injured party does not have to prove that the manufacturer was negligent. In some states, the legal doctrine of strict liability has been applied to developers who make mass-produced detached houses, apartments, or condominiums. The argument was made that the developer, like a manufacturer, is producing a product that has strong similarities from one unit to another. Although the shape or appearance of the individual homes may be different, all have essentially the same elements, such as concrete foundations and wood stud walls. Thus, a developer mass-produces a product, like a manufacturer, and can be held strictly liable for defects.

Certificate of merit.   In California, an attorney must follow Sec. 411.35 of the *Code of Civil Procedure* in order to sue a third-party design professional. This code section requires the attorney to file a certificate of merit with the court prior to naming a design professional as a defendant in a complaint.

The attorney who sues a design professional engineer must perform the following actions in order to file a certificate of merit with the court:

1. Review the facts and main issues of the case.

2. Consult with a registered engineer who is not a party to the lawsuit.

3. Conclude on the basis of the review of the case and the consultation with the registered engineer that "there is reasonable and meritorious cause for the filing of (the) action."

After the certificate of merit has been filed with the court, the attorney can then serve the design professional with a summons indicating that he or she is also a defendant in the lawsuit.

Declaration.   During the process of civil litigation, the forensic engineer may be asked by the attorney to prepare a declaration. This is a legal document that is in support or response to an attorney's motion brought before the court. The declaration should be prepared jointly by the forensic engineer and attorney, and must be signed under penalty of perjury by the forensic engineer.

There can be many different situations where the forensic engineer is asked to prepare a declaration. For example, a lawsuit was initiated by an owner who experienced damage when a drain line ruptured during a heavy rainstorm. The city municipality, a cross-defendant in the lawsuit, filed a motion with the court asking that it be released from the lawsuit. As part of the motion for dismissal, the city's forensic engineer prepared a declaration stating that based on his visual observations, the drain line was not on city property. The owner's forensic engineer performed a land survey, and countered with a declaration stating that based on the land survey, a portion of the drain line was indeed on the city's property. The judge reviewed the motion and the declarations by the forensic engineers and ruled that the city could not be dismissed from the lawsuit.

Case management order. The presiding judge in a lawsuit may issue a case management order. Appendix B presents an example of a case management order. This document contains the judge's instructions on how the lawsuit will proceed. The case management order may also specify deadlines for the completion of specific tasks, such as the deadlines for discovery and the taking of depositions. The case management order may also specify the date of trial. Exhibit H in App. B presents a case management summary of the deadlines for the lawsuit.

1.4.3 Discovery

Discovery is the process of obtaining information that is pertinent to the lawsuit. The judge will decide if the information will be admissible evidence at trial. Methods of discovery include interviewing of witnesses, researching public documents, and investigating documents, land, or other property. Two other methods of discovery are as follows:

Interrogatory. An interrogatory is a written question that is addressed to a specific person in the lawsuit. Exhibit A in App. B presents examples of interrogatories.

Miller (1993) states that interrogatories are a useful discovery tool because they can accomplish the following: (1) testing the merit of a claim or defense, (2) eliciting information that may lead to further evidence, (3) determining facts useful in planning further discovery, (4) particularizing vague or uncertain pleadings, (5) narrowing issues for trial, and (6) eliciting information that will help impeach the credibility of a witness in deposition or trial.

Deposition. The purpose of a deposition is to obtain the sworn testimony of a witness or forensic engineer prior to trial. Prior to a deposition, the witness will be subpoenaed. This legal document will indicate the date, location, and attorney requesting the deposition. If the subpoena says "duces tecum," then the person must bring all of his or her files to the deposition.

At the deposition, attorneys involved in the lawsuit will ask the witness specific questions concerning the project. A court reporter will then produce a stenographic transcript of the questions and answers, which is considered the official record of the deposition. Since this testimony is taken under oath, it is important that the opinions of the forensic engineer are clear and concise. The testimony of the forensic engineer at the time of trial must be consistent with the statements made during the deposition.

1.4.4   Alternative dispute resolution

Mediation. The parties involved with the civil litigation may agree on mediation before trial. In mediation, a retired judge, independent attorney, or special master will listen to evidence in an informal atmosphere and attempt to settle the case through compromise. At a mediation, the forensic engineer may be required to make a detailed presentation to the mediator, with cross-examination being, hopefully, restrained by the mediator. It is in the mediator's interest to settle the case, not have the forensic engineer discredited or impeached.

Arbitration. An arbitration is more formal than a mediation. In some cases, such as insurance-related litigation, the policy may state that the dispute can be settled only through binding arbitration. In other cases, the parties in a lawsuit enter into an agreement (contract) to arbitrate the case and specify certain procedures that will govern the process, such as using the format established by the American Arbitration Association (AAA). At an arbitration, the forensic engineer will give testimony, under oath, to an arbitrator or arbitration panel that will decide the case. Arbitration is usually less expensive than a jury trial and it can be completed in less time. Another advantage of arbitration versus a jury trial is that the parties involved in the dispute can select arbitrators who are experienced with construction defect law.

1.4.5   Trial

The forensic engineer should be prepared to present the results of the investigation in a court of law. There are publications that help the individual prepare for trial testimony as well as describe the proper conduct during trial. For example ASCE, in *Guidelines for Failure Investigation* (Greenspan et al., 1989), lists several important factors when giving testimony, such as presenting a positive attitude and proper demeanor, showing respect to the court, and maintaining your professionalism and proper posture. Other important items during testimony are to speak clearly, pause before answering the question (to allow time for objections), and only answer the question, without expanding on your answers, which could give the opposing attorney more information than is required. The following is a summary of important items to bear in mind when giving testimony (Greenspan et al., 1989):

- *Attitude and demeanor:* Present a positive attitude and proper demeanor.

- *Respect:* Show proper respect to the court. Address the judge as "your honor."

- *Qualifications:* Do not be boastful or exaggerate professional qualifications.

- *Body language:* Remain calm and be aware of body language.

- *Posture:* Sit up straight in the witness stand.

- *Gestures:* Avoid excessive hand or head gestures.

- *Eye Contact:* Look directly at the jury when answering questions.

- *Facial expressions:* Avoid facial expressions that convey feelings.

- *Speak clearly:* Speak concisely in a clear voice.

- *Pause before answering:* A momentary pause allows time for attorneys to voice objections.

- *Remain dispassionate:* Maintain composure during verbal attacks by the opposing attorneys.

- *Avoid confrontation:* Confrontation with opposing attorneys damages the expert's credibility.

The most difficult time during trial testimony is usually during cross-examination, when opposing attorneys may try to discredit your testimony. Matson (1994) states:

> The opposing attorney has the opportunity to question the validity of your opinions expressed during direct examination. They will also question the

veracity of the witness—*you*. This is the opportunity the opposing attorney has been waiting for. The strategy is to impeach your testimony and destroy your credibility. They want to make you sound unbelievable to the jury. The attorney has spent much of their professional life training to discredit you in cross examination...."here are the facts; debate them," is the *modus operandi* for the lawyer.

1.5   Examples

This section provides two examples of earth movement that occurred during the "El Niño" winter of 1997–1998. Both examples involve landslides that were triggered by the heavy California rainfall due to the El Niño weather pattern. Rainfall can infiltrate into the ground where it can trigger landslides by lubricating slide planes and by raising the groundwater table, which increases driving forces due to seepage pressures and decreases resisting forces due to the buoyancy effect.

Figures 1.1 to 1.7 show pictures of the first landslide which occurred in Olivenhain, California, in March 1998. The figures show the following:

- *Figures 1.1 and 1.2.*   These two photographs show the toe of the landslide. The smaller arrow in Fig. 1.1 points to groundwater which was observed flowing out of the toe of the landslide. The two larger arrows in Fig. 1.1 point to areas where the toe of the landslide is overriding and rolling down the slope. Figure 1.2 is a close-up view that shows the toe of the landslide rolling down the slope.

- *Figures 1.3 and 1.4.*   These two photographs show ground cracks and distress at the main body of the landslide. In Fig. 1.3, the force of the landslide has crushed a gunite drainage ditch. Figure 1.4 shows ground cracks and distortion of a gunite drainage ditch at another location.

- *Figures 1.5 to 1.7.*   These three photographs show the head of the landslide and the main scarp. This area is at the top of the landslide and the mass movement of the landslide has caused the ground to drop down. In Fig. 1.5, the main scarp runs underneath the wood deck, which has been displaced downward and laterally. The arrow in Fig. 1.5 points to the main scarp, and Fig. 1.6 is a close-up view of this area. Another view of the main scarp is shown in Fig. 1.7, where the ground surface has dropped downward by about 2 m (7 ft). The vertical distance between the two arrows in Fig. 1.7 is the amount of vertical ground displacement caused by movement of the landslide. Note in Fig. 1.7 the slick and grooved

Figure 1.1 Olivenhain Landslide: Toe of landslide. (Note: the smaller arrow points to groundwater observed coming out of the landslide and the larger arrows point to the toe of the landslide rolling downslope.)

Figure 1.2 Olivenhain Landslide: Close-up view of the toe of the landslide rolling downslope.

Figure 1.3 Olivenhain Landslide: Main body of the landslide which has crushed a gunite drainage ditch.

Figure 1.4 Olivenhain Landslide: Another view of the main body of the landslide, showing ground cracks and damage to another drainage ditch.

Figure 1.5 Olivenhain Landslide: View at the head of the landslide. (Note: arrow points to the main scarp.)

Figure 1.6 Olivenhain Landslide: Close-up view of main scarp indicated in Figure 1.5.

Figure 1.7 Olivenhain Landslide: Another view of the main scarp. (Note: the vertical distance between arrows indicates the amount that the ground has dropped down.)

nature (i.e., slickensides) of the material exposed on the main scarp.

Figures 1.8 to 1.16 show pictures of a second landslide which occurred in Laguna Niguel, California, in March 1998. This landslide is much larger and deeper and caused more extensive damage. The mass movement of the landslide caused a dropping down at the top (head) and a bulging upward of the ground at the base (toe) of the landslide. The figures show the following:

- *Figures 1.8 to 1.10.*   These photographs show damage to condominiums at the toe of the landslide. The area shown in Fig. 1.8 was originally relatively level, but the toe of the landslide has uplifted both the road and the condominiums. The arrow in Fig. 1.8 indicates the original level of the road and condominiums. Note also in Fig. 1.8 that the upward thrust caused by the toe of the landslide has literally ripped the building in half. At another location, the toe of the landslide uplifted the road by about 6 m (20 ft), as shown in Fig. 1.9. Many buildings located at the toe of the landslide were completely crushed by the force of the landslide movement (Fig. 1.10).

Figure 1.8 Laguna Niguel Landslide: View of the toe of the
landslide. (Note that the arrow indicates the original level
of this area, but the landslide uplifted both a portion of the
road and the condominium.)

- *Figures 1.11 to 1.16.*   These photographs show damage at the head
  of the landslide. The main scarp, which is shown in Fig. 1.11, is
  about 12 m (40 ft) in height. The vertical distance between the two
  arrows in Fig. 1.11 represents the down-dropping of the top of the
  landslide. Figure 1.12 shows a corner of a house that is suspended
  in midair because the landslide has dropped down and away.
  Figure 1.13 shows a different house where the rear is also suspended in midair because of the landslide movement. A common
  feature of all the houses located at the top of the landslide was the
  vertical and lateral movement caused by the landslide as it

Figure 1.9 Laguna Niguel Landslide: View of the toe of the landslide where the road has been uplifted 20 feet.

dropped down and away. For example, Fig. 1.14 shows one house that has dropped down relative to the driveway, Fig. 1.15 shows a second house where a gap opened up in the driveway as a house was pulled downslope, and Fig. 1.16 shows a third house (visible in the background) that was pulled downslope, leaving behind a groove in the ground, which was caused by the house footings being dragged across the ground surface.

As these two examples illustrate, landslides can be some of the most destructive failures investigated by the forensic engineer. The tools needed by the forensic engineer to investigate landslides and the methods to stabilize landslides will be discussed in Chaps. 6 and 9.

1.6 Outline of Chapters

The book is divided into four separate parts. Part 1 (Chaps. 2 and 3) provides the basic procedures that can be applied to geotechnical and foundation engineering forensic investigations. Part 2 (Chaps. 4 to 8) deals with specific problems typically encountered by the forensic engineer, such as settlement, expansive soil, slope movement, and moisture intrusion problems. Part 3 (Chap. 9) provides a discussion of

Figure 1.10 Laguna Niguel Landslide: View of damage at the toe of the landslide. The landslide has crushed the buildings located in this area.

the forensic engineer's role in the development of repairs, and Part 4 (Chap. 10) is a concluding chapter with a presentation of strategies for avoiding civil liability. Throughout the book, case studies are presented to illustrate the diverse nature of geotechnical and foundation problems that can be encountered by the forensic engineer.

A list of symbols is provided at the beginning of some chapters. An attempt has been made to select those symbols most frequently listed in standard textbooks and used in practice. In terms of units, both the International System of Units (SI) and the United States Customary System units (USCS) are provided.

Figure 1.11 Laguna Niguel Landslide: Head of the landslide. (Note: the vertical distance between the two arrows is the distance that the landslide dropped downward.)

Figure 1.12   Laguna Niguel Landslide: Head of the land-slide. (Note: the smaller arrow points to the corner of a house which is suspended in midair and the larger arrow points to the main scarp.)

Figure 1.13 Laguna Niguel Landslide: Head of the landslide showing another house with a portion suspended in midair. (Note: the arrow points to a column and footing that are suspended in midair.)

Figure 1.14 Laguna Niguel Landslide: Head of the landslide where a house has dropped downward. (Note: the arrow points to the area where the house foundation has punched through the driveway on its way down.)

Figure 1.15 Laguna Niguel Landslide: Head of the landslide showing lateral movement of the entire house and a portion of the driveway. (The distance between the arrows indicates the amount of lateral pulling of the house.)

Figure 1.16 Laguna Niguel Landslide: Head of the land-slide showing a house that has been pulled downslope. (Note: the groove in the ground was caused by the house footings being dragged across the ground surface.)

Assignment
and Investigation

Two views of the Neptune Bluff failure. The arrow in the upper photograph points to a collapsed house. The lower photograph shows a view from the top of the sea cliff and provides a close-up view of the collapsed house.

2

The Assignment

2.1 Preliminary Information

As with any project, basic information should be obtained before accepting the assignment. This includes data on the client and the location and type of project. In forensic engineering, it would also be important to obtain a preliminary idea of the nature of the structural damage or deterioration at the site. A list of preliminary information that should be obtained is provided below:

1. *Type of forensic project.* It is important to determine whether the project deals directly with civil litigation or involves a nonlitigation forensic investigation for an insurance company, independent adjuster, property owner, or governmental agency.

2. *Nature of damage or deterioration.* The type and extent of damage to the structure should be obtained from the client. The project should be accepted only if it is within your area of expertise.

3. *Age of the problem.* Determine whether the damage is a recent problem or developed some time in the past. If the problem developed some time in the past, ask if a forensic engineer had previously been retained and what happened to the expert. Request the expert's name and make contact in order to determine why he or she is no longer involved.

4. *Scope of work.* In many cases, the entire scope of the investigation is difficult to determine at the outset. It is usually best to agree on a limited initial scope, such as a preliminary site visit. After the preliminary site visit, a plan of investigation and cost estimate could then be developed.

5. *Conflict of interest.*   Do not work on the project if there is a conflict of interest. A conflict of interest could include parties involved with the project as well as any work that you have performed on the project during its construction or past repair or rehabilitation.

2.2   Accepting the Assignment

If the assignment deals with civil litigation, it is important to obtain as much information about the lawsuit as possible. For example, the forensic engineer must know whether the client is the plaintiff, defendant, or one of the cross-defendants. In some situations, the cross-defendants will jointly agree to hire the same forensic engineer and then share in the costs. Whatever the case, the forensic engineer must have a clear understanding of the clients involved in the lawsuit.

Shuirman and Slosson (1992) have developed a routine checklist that should be completed before accepting an assignment involving civil litigation. The checklist is reproduced below:

1. Be certain there are no conflicts of interest with any of the parties connected with the lawsuit, directly or indirectly.

2. Obtain as much background information on the disaster or failure as possible, mindful that such information may be biased or lack pertinent data.

3. Attempt to separate facts from opinions so as to be able to form an objective early picture of the main issues.

4. Inquire as to the status of the case and its tentative schedule to determine whether there is sufficient time for a thorough investigation.

5. Make certain the subject matter is within the appropriate areas of expertise for you or your firm. The attorney should clearly understand which issues are within that province and which are not.

6. Discuss fee schedules and determine when and by whom payments are to be made.

7. Check out the reputation of the attorney or law firm if it is not already known.

2.2.1   Curriculum vitae

In addition to the above checklist, the attorney should be informed of the forensic engineer's educational background, past work experience, state registrations, and publications. This can be accomplished by sending the attorney a curriculum vitae or résumé which lists your qualifications as a forensic engineer. Because this document can be used at trial to demonstrate that you are an expert, it should have a clear and concise format, be well organized, and be as complete as possible. A list of the items that should be included in the curriculum vitae are as follows (Greenspan et al., 1989):

1. *Identification.*   List your name, title, and business address.

2. *Profession.*   Indicate your title within the organization and the area of specialization (e.g., President, John Doe and Associates, Consulting Geotechnical Engineer).

3. *Education.*   Specify university, type of degree, and date. Indicate if degree was awarded with honors.

4. *Experience.*   Summarize job experience since graduation. List employer, dates, job title and a brief summary of your experience.

5. *Professional registration.*   Specify those states where you are registered as well as registration numbers.

6. *Published works.*   List in chronological order all published works and format this data as a bibliography, stating coauthors, year of publication, title, and publisher.

7. *Professional society membership and special appointments.*   List all memberships in professional societies and delineate type of membership (member, fellow, etc.). Also list special appointments, such as engineering state board member, etc.

8. *Awards.*   Indicate all awards relating to your profession, such as design awards, professional society awards, etc.

9. *Representative projects.*   In some cases, it may be helpful to include an overview of your design and investigative experience. For example, a list of about 10 projects could be included to demonstrate a wide range of expertise. It is best to combine different types of projects in order to obtain a balance between forensic assignments and design projects.

It is important that your curriculum vitae or résumé does not contain any false or misleading information. It is likely that opposing attorneys will investigate the items listed on your curriculum vitae to determine their accuracy. The author is aware of a case where a forensic engineer listed a university degree that he did not have. In front of the judge and jury, the forensic engineer was confronted with the fact that the information on his curriculum vitae was fraudulent.

Of course, the engineer's credibility was destroyed at the trial and in addition, the state board of registration suspended his engineering license. The most important items for a curriculum vitae are that it is completely accurate, comprehensive, and updated to reflect current conditions.

2.2.2 Compensation

The appropriate type of compensation is an hourly fee or per diem rate for services, with a higher rate for deposition or trial testimony. It is best to send the client a fee schedule, which lists the hourly or per diem rates. It is usually difficult to accurately predict the cost of a complete forensic investigation. The best approach is to provide the client with a first-phase cost estimate, which would include an initial site visit. After the initial site visit, an approximate cost of the proposed investigation could then be developed.

In no case may a forensic engineer accept a project on the basis of a contingency fee or percentage of the monetary damage award. While it is standard for the plaintiff's attorney to accept a case where the fee is a certain percentage of the amount awarded by the jury, it is unethical for the forensic engineer to pursue such an agreement. This is because the forensic engineer must be unbiased and the outcome of the case must have no bearing on the forensic engineer's compensation.

2.2.3 Forensic engineer's agreement with the client

Before work progresses on the project, it is important to have the client sign a forensic engineering agreement, which is basically a contract between the forensic engineer and the client. The agreement should include such items as the agreed-upon hourly fee, clarification of services, invoicing and payment procedures, and protection of your work product. Some clients, such as insurance companies, may refuse to sign the forensic engineering agreement. In these cases, a letter of authorization for forensic services may serve as the contract between the insurance company and the forensic engineer.

2.2.4 Confidentiality

Especially in forensic investigations that involve litigation, it is essential that all information be strictly confidential. Attorneys may have developed a preliminary strategy for dealing with the case, and this strategy could be destroyed if confidential information is obtained by opposing attorneys.

At some point in the litigation, such as at your deposition, all the information in your file may become discoverable. At that time the information will no longer be confidential, and opposing attorneys will be able to question you on the file contents. For a forensic investigation involving litigation, it is important that the forensic engineer remember that all the contents in his or her file may eventually be discoverable. Inappropriate material, such as rough drafts of reports and letters, personal documents, and inappropriate or unrelated material should not be placed in the file.

3

The Investigation

3.1 Planning the Investigation

This chapter describes the steps necessary to complete a forensic investigation with the ultimate goal of determining the cause of failure. There are many excellent publications on the procedure and methods to perform a forensic investigation (e.g., Leonards, 1982; Carper, 1986). The elements of the investigation as described in this chapter are general in nature and they may not be inclusive for all types of investigations. Figure 3.1 (from Greenspan et al., 1989) shows the typical steps needed to complete a forensic investigation. As indicated in Fig. 3.1, the investigation often culminates with the preparation of a final report that indicates the cause of the failure, and in many cases, presents repair recommendations.

The first step in the actual investigation process should be to prepare an investigative plan. For the investigation of a minor problem, the planning effort may be minimal. But for those investigations of large-scale problems, the investigative plan can be quite extensive and could change as the investigation progresses. The planning effort should include the following (Greenspan et al., 1989):

- Budget and scheduling considerations
- Selection of an interdisciplinary team that will best respond to the failure situation
- Site observations and testing requirements
- Document collection
- Analysis and synthesis of data
- Development of failure hypothesis

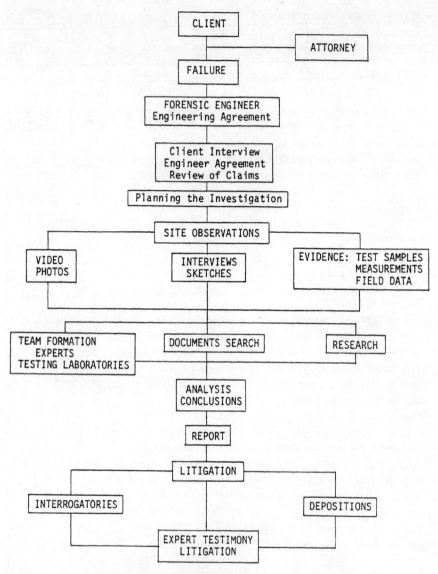

Figure 3.1 Forensic investigation. (*From Greenspan et al., 1989, reprinted with permission from the American Society of Civil Engineers.*)

Appendix B presents a case management order, issued by the judge for a project dealing with civil litigation. The case management order is important because it specifies the deadlines for completing the discovery process. For example, as indicated by the Summary of Significant Deadlines (Exhibit H in App. B), both the plaintiff's and defendant's destructive testing must be completed by certain dates. Likewise the dates for mediations, discovery cutoff, depositions, and trial are also set by the judge. The forensic engineer must be aware of these time limits when planning the investigation. It is likely that any evidence discovered after the discovery cutoff deadline will not be admissible evidence at the time of trial. This could have serious consequences on the outcome of the trial and the forensic engineer should always ask about court-ordered deadlines. The forensic engineer must plan the investigation so that it meets the court-ordered deadlines.

3.2   Site Investigation

3.2.1   Initial site visit

The purpose of the initial site visit is to evaluate the scope and nature of the failure. The initial site visit should be performed by the forensic engineer, who may be accompanied by assistants. For those structures having sudden damage or collapse, it is important to perform the initial site visit immediately after the assignment has been accepted. This is because evidence may be lost or disturbed as time goes by. At the initial site visit, it is important to take photographs of the observed damage and collect samples if it is likely that the samples will be lost or destroyed before the investigation can be completed.

Numerous photographs should be taken of the damaged structure. It may be appropriate to take a professional photographer who has the proper equipment for long-range and close-up photographs of the site. Sequentially numbering and marking the location of the photographs on a site plan sketch may help refresh recollection of where the pictures were taken from. In the excitement of the first visit, there is a natural tendency to forget about photographs.

For example, the author was involved in one project where an airport runway was damaged by a flood. The damage consisted of sand deposits on both sides of the runway cracks that were attributed to a loss of runway base material. Although eight engineers visited the site and observed the damage, no photographs or samples were taken of the sand deposits along the runway cracks. The next day, the maintenance crew swept the runway and all evidence of the sand deposits was lost. Without actual samples or photographs of the sand deposits, it was more difficult to convince the insurance company that the runway had been damaged during the flood.

3.2.2 Nondestructive testing

After the initial site visit, there are usually follow-up visits to prepare field sketches and field notes, conduct interviews, perform nondestructive testing, and install monitoring devices such as piezometers and inclinometers.

As the name implies, nondestructive testing does not cause any damage or disruption to the site. An example of nondestructive testing is geophysical techniques that can be used to locate underground voids or buried objects, such as oil tanks. Similar to geophysical techniques is acoustic emission, which uses high-frequency sound waves to detect flaws in engineering structures. For example, Kisters and Kearney (1991) state "Acoustic emission is presently being used to monitor crack propagation in bridges. This technique can be easily adapted to steel lock and dam structures." According to Rens et al. (1997), other nondestructive testing includes thermal, ultrasonic, and magnetic methods. Thermal techniques have been applied to several civil engineering projects such as asphalt and pavement condition assessments (Rens et al., 1997).

Another example of nondestructive testing is a manometer survey, which is also referred to as a *floor level survey*. It is a nondestructive means of finding flaws or design defects for foundations. A manometer survey is commonly used to determine the relative elevation of a concrete slab-on-grade or other foundation element. The survey consists of taking elevations at relatively close intervals throughout the interior floor slab. These elevation points are then contoured, much like a topographic map, to provide a graphic rendition of the deformation condition of the foundation. Soil movement can cause displacement of the foundation and the manometer survey can detect this deformation. For example, if one side or corner of a concrete slab is significantly lower than the rest of the foundation, this could indicate settlement or slope movement in that area. Likewise, if the center of the foundation is bulged upward, it would be detected by the manometer survey and this could indicate expansive soil. Other soil phenomena or slab conditions can also be detected by the manometer survey. For example, close deformation contours indicate a high angular distortion, which in many cases corresponds to the location of foundation cracks. Throughout the book, examples of manometer surveys that depict foundation displacement due to different soil mechanisms will be presented.

3.2.3 Destructive testing

Concerning destructive testing, Miller (1993) states:

> To document and cover the extent of resultant damages, it is often necessary to perform extensive destructive testing to determine the source and extent of the damage. Destructive testing can involve removing limited sections of the building so that access may be gained to concealed areas. If destructive testing is undertaken, it is imperative that it be performed in a logical and methodical fashion. Extensive documentation should be made during all portions of the destructive testing because it is a costly process, and repetition should be avoided if at all possible.

Subsurface exploration. An example of destructive testing is subsurface exploration, such as the excavation of test pits and borings. Table 3.1 [from Sowers and Royster (1978), based on the work by ASTM, Lambe (1951), Sanglerat (1972), and Sowers and Sowers (1970)] summarizes the boring, core drilling, sampling and other exploratory techniques that can be used by the forensic engineer. Figure 3.2 shows a bucket auger drilling rig which is routinely used to excavate a 0.6-m- (24-in.-) diameter boring that can then be downhole-logged by the geotechnical engineer or geologist. Figure 3.3 provides an example of the type of conditions that can be observed downhole. Figure 3.3a shows a knife that has been placed in an open fracture in bedrock. The open fracture in the rock was caused by massive landslide movement. Figure 3.3b is a side view of the same condition.

In many cases for forensic investigations, there will not be enough room to use the large drilling equipment shown in Fig. 3.2. For such projects having limited access, there are other types of drill rigs available to excavate borings. For example, Figs. 3.4 and 3.5 show two views of limited-access drill rigs. The tripod auger drill rig shown in Fig. 3.4 can be used in many confined areas, but has limited drilling power and can not excavate through cobbles, boulders, or hard rock, or to a great depth. Other types of limited-access drill rigs, such as that shown in Fig. 3.5, have more drilling power and can even excavate 0.6-m- (24-in.-) diameter borings.

Another example of limited access is the drilling of borings on slope faces. For example, Fig. 3.6 shows three views of the excavation of a boring on a steep slope face. The process began with the construction of a wood support platform on the slope face. Then the drill rig was lifted by a crane (Fig. 3.6a) and placed on the wood platform as shown in Fig. 3.6b. Once the drill rig was securely attached to the wood platform, the drilling process commenced as shown in Fig. 3.6c.

The test pits or borings are used to determine the thickness of soil and rock strata, estimate the depth to groundwater, obtain soil or

TABLE 3.1 Boring, Core Drilling, Sampling, and Other Exploratory Techniques

Method (1)	Procedure (2)	Type of sample (3)	Applications (4)	Limitations (5)
Auger boring, ASTM D 1452	Dry hole drilled with hand or power auger; samples preferably recovered from auger flutes	Auger cuttings, disturbed, ground up, partially dried from drill heat in hard materials	In soil and soft rock; to identify geologic units and water content above water table	Soil and rock stratification destroyed; sample mixed with water below the water table
Test boring, ASTM D 1586	Hole drilled with auger or rotary drill; at intervals samples taken 36 mm ID and 50 mm OD driven 0.45 m in three 150-mm increments by 64-kg hammer falling 0.76 m; hydrostatic balance of fluid maintained below water level	Intact but partially disturbed (number of hammer blows for second plus third increment of driving is standard penetration resistance or N)	To identify soil or soft rock; to determine water content; in classification tests and crude shear test of sample (N value: a crude index to density of cohesionless soil and undrained shear strength of cohesive soil)	Gaps between samples, 30 to 120 cm; sample too distorted for accurate shear and consolidation tests; sample limited by gravel; N value subject to variations depending on free fall of hammer
Test boring of large samples	50 to 75 mm ID and 63 to 89 mm OD samplers driven by hammers up to 160 kg	Intact but partially disturbed (no. of hammer blows for 2d plus 3d increment of driving is penetration resistance)	In gravelly soils	Sample limited by larger gravel
Test boring through hollow-stem auger	Hole advanced by hollow stem auger; soil sampled below auger as in test boring above	Intact but partially disturbed (no. of hammer blows for 2d plus 3d increment of driving is N value)	In gravelly soils (not well adapted to harder soils or soft rock)	Sample limited by larger gravel; maintaining hydrostatic balance in hole below water table is difficult
Rotary coring of soil or soft rock	Outer tube with teeth rotated; soil protected and held stationary inner tube; cuttings flushed upward by drill fluid (examples, Denison, Pitcher, and Acker samplers)	Relatively undisturbed sample, 50 to 200 mm wide and 0.3 to 1.5 m long in liner tube	In firm to stiff cohesive soils and soft but coherent rock	Sample may twist in soft clays; sampling loose sand below water table is difficult; success in gravel seldom occurs
Rotary coring of swelling clay, soft rock	Similar to rotary coring of rock; swelling core retained by third inner plastic liner	Soil cylinder 28.5 to 53.2 mm wide and 600 to 1500 mm long encased in plastic tube	In soils and soft rocks that swell or disintegrate rapidly in air (protected by plastic tube)	Sample smaller; equipment more complex
Rotary coring of rock, ASTM D 2113	Outer tube with diamond bit on lower end rotated to cut annular hole in rock; core protected by stationary inner tube; cuttings flushed upward by drill fluid	Rock cylinder 22 to 100 mm wide and as long as 6 m, depending on rock soundness	To obtain continuous core in sound rock (% of core recovered depends on fractures, rock variability, equipment, and driller skill)	Core lost in fracture or variable rock; blockage prevents drilling in badly fractured rock; dip of bedding and joint evident but not strike

Method	Procedure	Sample	Purpose	Remarks
Rotary coring of rock, oriented core	Similar to rotary coring of rock above; continuous grooves scribed on rock core with compass direction	Rock cylinder, typically 54 mm wide and 1.5 m long with compass orientation	To determine strike of joints and bedding	Method may not be effective in fractured rock
Rotary coring of rock, wire line	Outer tube with diamond bit on lower end rotated to cut annular hole in rock; core protected by stationary inner tube; cuttings flushed upward by drill fluid; core and stationary inner tube retrieved from outer core barrel by lifting device or "overshot" suspended on thin cable (wire line) through special large-diameter drill rods and outer core barrel	Rock cylinder 36.5 to 85 mm wide and 1.5 to 4.6 m long	To recover core better in fractured rock, which has less tendency for caving during core removal; to obtain much faster cycle of core recovery and resumption of drilling in deep holes	Same as ASTM D 2113 but to lesser degree
Rotary coring of rock, integral sampling method	22-mm hole drilled for length of proposed core; steel rod grouted into hole; core drilled around grouted rod with 100- to 150-mm rock coring drill (same as for ASTM D 2113)	Continuous core reinforced by grouted steel rod	To obtain continuous core in badly fractured, soft, or weathered rock in which recovery is low by ASTM D 2113	Grout may not adhere in some badly weathered rock; fractures sometimes cause drift of diamond bit and cutting rod
Thin-wall tube, ASTM D 1587	75- to 1250-mm thin-wall tube forced into soil with static force (or driven in soft rock); retention of sample helped by drilling mud	Relatively undisturbed sample, length 10 to 20 diameters	In soft to firm clays, short (5 diameter) samples of stiff cohesive soil, soft rock, and, with aid of drilling mud, firm to dense sands	Cutting edge wrinkled by gravel; samples lost in loose sand or very soft clay below water table; more disturbance occurs if driven with hammer
Thin-wall tube, fixed piston	75- to 1250-mm thin-wall tube, which has internal piston controlled by rod and keeps loose cuttings from tube, remains stationary while outer thin wall tube forced ahead into soil; sample in tube is held in tube by aid of piston	Relatively undisturbed sample, length 10 to 20 diameters	To minimize disturbance of very soft clays (drilling mud aids in holding samples in loose sand below water table)	Method is slow and cumbersome

TABLE 3.1 (*Continued*) Boring, Core Drilling, Sampling, and Other Exploratory Techniques

Method (1)	Procedure (2)	Type of sample (3)	Applications (4)	Limitations (5)
Swedish foil	Samples surrounded by thin strips of stainless steel, stored above cutter, to prevent contact of soil with tube as it is forced into soil	Continuous samples 50 mm wide and as long as 12 m	In soft, sensitive clays	Samples sometimes damaged by coarse sand and fine gravel
Dynamic sounding	Enlarged disposable point on end of rod driven by weight falling fixed distance in increments of 100 to 300 mm	None	To identify significant differences in soil strength or density	Misleading in gravel or loose saturated fine cohesionless soils
Static penetration	Enlarged cone, 36-mm diameter and 60° angle, forced into soil; force measured at regular intervals	None	To identify significant differences in soil strength or density; to identify soil by resistance of friction sleeve	Stopped by gravel or hard seams
Borehole camera	Inside of core hole viewed by circular photograph or scan	Visual representation	To examine stratification, fractures, and cavities in hole walls	Best above water table or when hole can be stabilized by clear water
Pits and trenches	Pit or trench excavated to expose soils and rocks	Chunks cut from walls of trench; size not limited	To determine structure of complex formations; to obtain samples of thin critical seams such as failure surface	Moving excavation equipment to site, stabilizing excavation walls, and controlling groundwater may be difficult
Rotary or cable tool well drill	Toothed cutter rotated or chisel bit pounded and churned	Ground	To penetrate boulders, coarse gravel; to identify hardness from drilling rates	Identifying soils or rocks difficult
Percussion drilling (jack hammer or air track)	Impact drill used; cuttings removed by compressed air	Rock dust	To locate rock, soft seams, or cavities in sound rock	Drill becomes plugged by wet soil

NOTE: Reprinted with permission from *Landslides: Analysis and Control*, Special Report 176. Copyright 1978 by the National Academy of Sciences. Courtesy of the National Academy Press, Washington, D.C. (Sowers and Royster, 1978).

Figure 3.2 Bucket auger drill rig.

Figure 3.3a Knife placed in an open fracture in bedrock caused by landslide movement. (*Photograph taken downhole in a large-diameter auger boring.*)

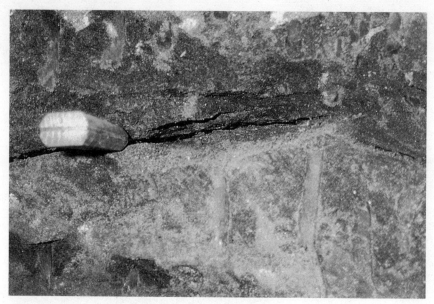

Figure 3.3b Side view of condition shown in Figure 3.3a.

Figure 3.4 Limited-access drilling equipment (arrow points to tripod drill rig).

Figure 3.5 Limited-access drill rig.

rock specimens, and perform field tests such as sand cone tests or standard penetration tests (SPTs). The Unified Soil Classification System (USCS) can be used to classify the soil exposed in the borings or test pits (Casagrande, 1948). The subsurface exploration and field sampling should be performed in accordance with standard procedures, such as those specified by the American Society for Testing and Materials (ASTM 1970, 1971, and 1997d) or other recognized sources (e.g., Hvorslev, 1949, and ASCE 1972, 1976, and 1978). An example of field exploration and testing is shown in Fig. 3.7, where a test pit has been excavated into an airport runway and a sand cone test is being performed on the runway base material.

Coring of foundations.   Another common type of destructive testing is the coring of foundations. By coring the foundation, the thickness of concrete, reinforcement condition, and any deterioration can be observed. Also, soil samples can be obtained directly beneath the foundation. Figure 3.8 shows a photograph of a slab-on-grade foundation that has been cored.

Other types of destructive testing.   There can be numerous other types of destructive testing performed by the forensic engineer. Examples

Figure 3.6*a*   Crane lifting drill rig (arrow points to drill rig).

include field load tests, cone penetration testing, and in-place testing, such as determining the shear strength of in-place soil or rock.

3.2.4   Monitoring

There are many types of monitoring devices used by forensic engineers. Some of the more common monitoring devices are as follows:

Inclinometers.   The horizontal movement preceding or during the movement of slopes can be investigated by successive surveys of the shape and position of flexible vertical casings installed in the ground (Terzaghi and Peck, 1967). The surveys are performed by lowering an inclinometer probe into the flexible vertical casing. The inclinometer probe is capable of measuring its deviation from the vertical. An ini-

Figure 3.6b Crane lowering drill rig onto wood platform.

Figure 3.6c Drilling operation on the face of a steep slope.

Figure 3.7 Test pit excavation.

Figure 3.8 Coring of a concrete slab-on-grade.

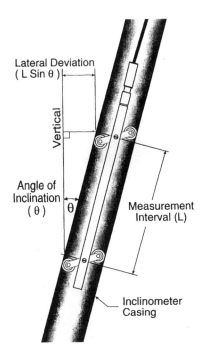

Lateral Deviation
(L Sin θ)

Vertical

Angle of
Inclination
(θ)   θ

Measurement
Interval (L)

Inclinometer
Casing

Figure 3.9 Inclinometer probe in
a casing. (*Reprinted with permission from the Slope Indicator
Co.*)

tial survey (base reading) is performed and then successive surveys
are compared to the base reading to obtain the horizontal movement
of the slope.

Figure 3.9 (from Slope Indicator, 1996) shows a sketch of the inclinometer probe in the casing and the calculations used to obtain the
lateral deformation. Throughout the book, examples of inclinometer
monitoring that depict different lateral movement conditions will be
presented.

Piezometers.   Piezometers are routinely installed in order to monitor
pore water pressures in the ground. Several different types are commercially available, including borehole, embankment, and push-in
piezometers. Figure 3.10 (from Slope Indicator, 1996) shows an example of a borehole piezometer.

In their simplest form, piezometers consist of a standpipe that can
be used to monitor groundwater levels and obtain groundwater samples. Figure 3.11 (from Slope Indicator, 1996) shows an example of a
standpipe piezometer.

Settlement monuments or cells.   Settlement monuments or settlement
cells can be used to monitor settlement or heave. Figure 3.12 (from
Slope Indicator, 1996) shows a diagram of the installation of a pneu-

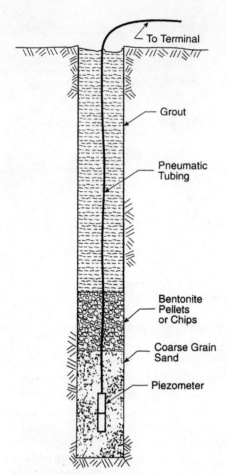

To Terminal

Grout

Pneumatic
Tubing

Bentonite
Pellets
or Chips

Coarse Grain
Sand

Piezometer

Figure 3.10 Pneumatic piezometer installed in a borehole. (*Reprinted with permission from the Slope Indicator Co.*)

matic settlement cell and plate. More advanced equipment includes settlement systems installed in borings that can not only measure total settlement, but also the incremental settlement at different depths.

Crack pins. A simple method to measure the widening of a concrete crack is to install crack pins on both sides of the crack. By periodically measuring the distance between the pins, the amount of opening or closing of the crack can be determined.

Other crack-monitoring devices are commercially available. For example, Fig. 3.13 shows an Avongard crack-monitoring device. There are two installation procedures: (1) the ends of the device are anchored by the use of bolts or screws or (2) the ends of the device are anchored with epoxy adhesive. The cen-

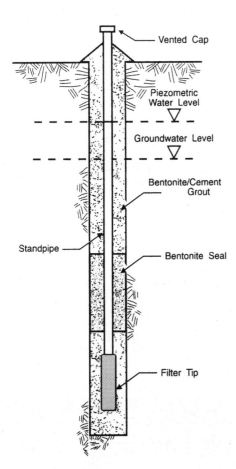

Vented Cap

Piezometric
Water Level

Groundwater Level

Bentonite/Cement
Grout

Standpipe

Bentonite Seal

Filter Tip

Figure 3.11 Standpipe (Casagrande) piezometer. (*Reprinted
with permission from the Slope
Indicator Co.*)

ter of the Avongard crack-monitoring device is held together with
clear tape, which is cut once the ends of the monitoring device have
been securely fastened with bolts, screws, or epoxy adhesive.

Other monitoring devices. There are many other types of monitoring
devices that can be used by the forensic engineer. Some commercially
available devices include pressure and load cells, borehole and tape
extensometers, soil strainmeters, beam sensors and tiltmeters, and
strain gauges.

3.2.5 Laboratory testing

Soil, groundwater, or foundation samples recovered from the site visits
can be returned to the laboratory for testing. Laboratory tests are commonly used to determine the classification, moisture and density, index

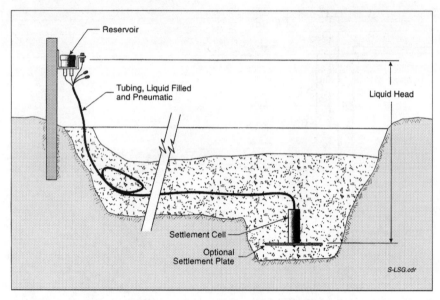

Figure 3.12 Pneumatic settlement cell installation. (*Reprinted with permission from the Slope Indicator Co.*)

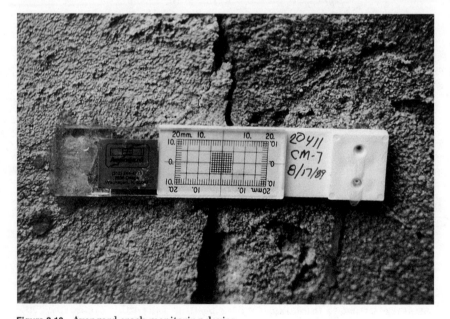

Figure 3.13 Avongard crack-monitoring device.

Figure 3.14 Shear failure of a saturated specimen of clay due to an increase in vertical pressure (arrows point to the failure surfaces).

properties, shear strength, compressibility, and hydraulic conductivity of the soil. For example, Fig. 3.14 shows the shear failure of a saturated specimen of clay due to an increase in vertical pressure. The arrows point to the failure surfaces. The shear strength of soil is important because the majority of slope instability and bearing capacity failures involve the shear rupture of the soil. Table 3.2 presents a summary of common soil laboratory tests used in forensic investigations.

Usually at the time of the laboratory testing, the forensic engineer will have developed one or more hypotheses concerning the cause of failure. The objective of the laboratory testing should be to further investigate these failure hypotheses. It is important that the forensic engineer develop a logical laboratory testing program with this objec-

TABLE 3.2   Common Soil Laboratory Tests Used in Forensic Investigations

Type of problem (1)	Soil properties (2)	Specification (3)
General	Classification	ASTM D 2487-93
	Particle size	ASTM D 422-90
	Atterberg limits	ASTM D 4318-95
	Water (or moisture) content	ASTM D 2216-92
	Wet density	Block samples or sampling tubes
	Specific gravity	ASTM D 854-92
Settlement (Chap. 4)	Consolidation	ASTM D 2435-96
	Collapse	ASTM D 5333-96
	Organic content	ASTM D 2974-95
	Fill investigation: standard proctor	ASTM D 698-91
	Fill investigation: modified proctor	ASTM D 1557-91
Expansive soil (Chap. 5)	Swell	ASTM D 4546-96
	Expansion index	ASTM D 4829-95 or UBC 18-2
Shear strength for slope movement (Chap. 6)	Unconfined compressive strength	ASTM D 2166-91
	Unconsolidated undrained	ASTM D 2850-95
	Consolidated undrained	ASTM D 4767-95
	Direct shear	ASTM D 3080-90
	Ring shear	Stark and Eid, 1994
	Miniature vane	ASTM D 4648-94
Erosion (Chap. 7)	Dispersive clay	ASTM D 4647-93
	Erosion potential	Day, 1990b
Deterioration (Chap. 7)	Pavements: CBR	ASTM D 1883-94
	Pavements: R value	ASTM D 2844-94
	Sulfate	Chemical analysis
Permeability (Chap. 8)	Constant head	ASTM D 2434-94
	Falling head	ASTM D 5084-90

tive in mind. The laboratory tests should be performed in accordance with standard procedures, such as those recommended by the American Society for Testing and Materials (ASTM) or those procedures listed in standard textbooks or specification manuals (e.g., Lambe, 1951; Bishop and Henkel, 1962; Department of the Army, 1970; *Standard Specifications*, 1997).

For some geotechnical forensic investigations, it may be important to determine the potential for future soil movement and damage. In these cases, the laboratory testing should model future expected conditions so that the amount of movement or stability of the ground can be analyzed.

For forensic investigations involving civil litigation, samples that are not irrecoverably damaged or destroyed during the laboratory testing should be saved and preserved so that they do not become contaminated. This is because other forensic experts involved in the case may want to observe or test the specimens. Also, at the time of trial, the specimens may need to be admitted as evidence.

3.3   Document Search

Table 3.3 presents a list of typical documents that may need to be reviewed for a forensic investigation. The following is a brief summary of these types of documents:

3.3.1   Project reports and plans

The reports and plans that were generated during the design and construction of the project may need to be reviewed. The reports and plans can provide specific information on the history, design, and construction of the project. These documents may also provide information on maintenance or alterations at the site.

As part of the discovery process for projects dealing with civil litigation, attorneys commonly subpoena or the judge may order that the entire records of the owner, designers, and contractors be placed in a document depository. Once the documents are submitted to the document depository, they will be available to all parties involved in the lawsuit.

The judge assigned the case will normally issue an order detailing the procedures to be followed concerning the document depository. For example, in the case management order (App. B, item 9) the judge has ordered that all parties in the lawsuit must submit the documents to the depository by a specific date. Exhibits C and D in App. B provide examples of documents to be deposited by the plaintiffs and defendants. Failure to comply with the judge's order can result in sanctions.

Once documents have been submitted to the document depository, the depository coordinator will place a "bate stamp" (which is a prefix

TABLE 3.3   **Typical Documents That May Need to Be Reviewed for a Forensic Investigation**

Project phase (1)	Type of documents (2)
Design	Design reports, such as geotechnical reports, planning reports, feasibility studies
	Design calculations and analyses
	Computer programs used for the design of the project
	Design specifications
	Applicable building codes
	Shop drawings and design plans
Construction	Construction reports, such as inspection reports, field memos, laboratory test reports, mill certificates
	Contract documents (contract agreements, provisions, etc.)
	Construction specifications
	Project payment data or certificates
	Field change orders
	Information bulletins used during construction
	Project correspondence between different parties
	As-built drawings (such as as-built grading plans, foundation plans)
	Photographs or videos
	Building department permits and certificate of occupancy
Postconstruction	Postconstruction reports, such as maintenance reports, modification documents, reports on specific problems, repair reports
	Photographs or videos
Technical data	Available records such as weather reports, seismic activity
	Reference materials, such as geologic and topographic maps, aerial photographs
	Technical publications, such as journal articles that describe similar failures

code and number) on each page or map. Exhibit E in App. B provides an example of bate stamp prefix codes. The bate stamp is named after the person who invented the sequential numbering apparatus. Many documents in the depository will be irrelevant in terms of the cause of the failure, and the forensic engineer must be able to distinguish the important applicable documents from the useless data. For example,

Matson (1994) describes a case where the opponents flooded the document depository with irrelevant papers:

> The approach was the "needle in the haystack" in which the opponents supplied tons of paper so overwhelming that my eyes became bloodshot reading only box labels. I spent all my time searching for meaningful documents in the flood of paper.

3.3.2 Building codes

A copy of the applicable building code at the time of construction should be reviewed. It has been argued that the standard of care is simply to perform work in accordance with the local building code. While performing work in accordance with building codes may reduce potential liability, it is still possible that in a court of law, a design engineer will be held to a higher standard. For example, Shuirman and Slosson (1992) state that, in many jurisdictions, the building codes and code enforcement may not meet current professional standards, and design engineers cannot rely on building codes to indemnify them from liability.

3.3.3 Technical documents

During the course of the forensic investigation, it may be necessary to check reference materials, such as geologic maps or aerial photographs. Other useful technical documents can include journal articles that may describe a failure similar to the one under investigation. For example, the ASCE *Journal of Performance of Constructed Facilities* deals specifically with constructed-related failures or deterioration.

3.4 Analysis and Conclusions

For forensic investigations, all of the data obtained from the investigation must be analyzed and the cause of the failure (failure theory) must be determined. Calculations or computer analyses may be needed to help analyze different failure hypotheses or the potential for future damage. It may even be appropriate to model the failure by using, for example, finite element analyses (e.g., Poh et al., 1997; Duncan, 1996). A thorough analysis is especially important with projects involving lawsuits because one objective is usually to determine proportional responsibility. On the basis of the cause of the failure, the forensic engineer will be able to offer opinions on who is responsible for the failure and proportion the responsibility between different parties.

In performing the analysis and in the development of conclusions, the forensic engineer may need to rely on the expertise of other forensic specialists. For example, geological studies are often essential in the investigation of landslides, rockfall, and seismic activity (Norris and Webb, 1990).

There can be many different causes of failure. Some of the more common causes of failure are related to inadequate subsurface exploration and laboratory testing, technical deficiencies or design errors, specification mistakes, improper construction, and defective materials (Greenspan et al., 1989). The failure theory must be well thought out and based on the facts of the case. It is not unusual that the forensic engineers involved in a case will disagree on the cause of the failure. Common reasons for a disagreement on the cause of the failure include the loss of evidence during the failure, the presence of conflicting test data, or substantially different eyewitness accounts of the failure. The following case study provides an example of differing failure theories.

3.4.1 Case study

The case study deals with the failure of a levee located on the north bank of the King's River in Fresno, California. The levee broke on April 10, 1995. Figure 3.15 shows a photograph taken during the levee break. A total length of about 60 m (200 ft) of the levee was eventually washed out. Because of the levee break, there was extensive flooding of the surrounding area as shown on the left side of Fig. 3.15. The flood waters caused several million dollars in damage, including crop losses and actual flooding of homes and business. Fortunately, there was no loss of life during the flood.

The damaged homeowners and farmers (plaintiffs) joined together in a lawsuit against two main defendants: the maintenance district and the irrigation district. The maintenance district's responsibility was to maintain and repair the levees as well as remove any obstructions (tree and vegetation growth) from the river channel. The irrigation district was responsible for pumping the seepage that came through the levee back into the river. There were a series of ditches that channeled the

Figure 3.15   April 10, 1995, levee break.

seepage to collection ponds, and then the water was pumped back into the river. The case settled out of court in September 1997.

Three different forensic geotechnical engineers where hired for the project: the plaintiff's expert, the maintenance district's expert, and the irrigation district's expert. The author was the irrigation district's expert. Three substantially different failure theories were developed as to the cause of the problem, as follows:

Plaintiff's expert. The plaintiff's expert concluded the following:

> We conclude that the cause of the failure was seepage accelerating to piping and ultimately failure under the increased head differential resulting from the greater flows in the levee. Our conclusion is based in part on the boring log [Fig. 3.16] in the vicinity of the failure which indicates that the levee and foundation materials are relatively pervious, highly susceptible

BORING LOG

Project/Client: KING'S RIVER LEVEE   F.N.:   N/A

Location: Top of Levee   Date: 1957

Estimated Surface Elevation (ft): _____ Total depth (ft): 28 Rig Type _____

Depth (Feet)	Sample Type	Sample Depth	Blow Count	Field Description By:
				Surface Conditions: Level, top of levee.
				Subsurface Conditions: FORMATION: Classification, color, moisture, tightness, etc.
0				From 0-5.0', Silty Sand (SM), non-plastic, 16 percent passing No. 200 sieve,
1				moisture content = 2 percent.
2				
3				
4				
5				From 5.0'-10.0, Sand (SP), poorly graded, non-plastic, 8 percent passing No. 200 sieve,
6				moisture content = 4 percent.
7				
8				
9				
10				From 10.0'-14.0', Silty Sand (SM), non-plastic, 41 percent passing No. 200 sieve,
11				moisture content = 20 percent.
12				
13				
14				From 14.0-15.0', Peat (Pt), 100 percent passing No. 200 sieve, moisture content = 102 percent.
15				From 15.0'-25.0', Silty Sand (SM), non-plastic, 15 percent passing No. 200 sieve,
16				moisture content = 19 percent.
17				
18				
19				
20				
21				
22				
23				
24				
25				From 25.0'-28.0', Sand (SP), poorly graded, non-plastic, 5 percent passing No. 200 sieve,
26				moisture content = 20 percent.
27				
28				Total Depth = 28 feet
29				Water at 10 feet.
30				

Figure 3.16 Boring log at top of levee (1957).

to seepage and piping and consist primarily of clean sands and slightly silty sands, and the reported previous and ongoing seepage problems at the failure site.

We considered the following alternatives to explain why the levee failed on April 10, 1995, and not previously during equal or greater flows: (1) the water level in the levee was higher than ever before and resulted in a greater head differential and hence greater seepage/piping velocity; (2) between the previous high flow year of 1986 and the 1995 flows, ongoing seepage ultimately reduced the integrity of the levee to a point where the flows of 1995 triggered the failure; and (3) the high flows of 1995 extended for a prolonged period of time and resulted in sufficient accelerated seepage to cause piping and ultimately failure.

The plaintiff's expert also stated:

Sloughing was apparently also ongoing at the specific site of the levee failure as documented by a photograph showing [the maintenance district's] stakes indicative of monitoring of sloughing. The pattern of the four stakes tied together with flagging was reduced to three stakes by the time of the photograph (April 10, 1995).

The plaintiff's expert concluded that the maintenance district was responsible for the failure because it was the district's job to maintain the levee and the district was monitoring the condition of the levee at the time of failure. The plaintiff's expert had no opinions concerning the responsibility of the irrigation district.

Maintenance district's expert. The maintenance district's expert testified at his deposition that the cause of the levee break was due to hydraulic fracturing. This is a process by which water pressure opens cracks in the ground which leads to sudden failure of the levee. The maintenance district's expert also testified that the irrigation district was partly responsible for the failure because they had excavated drainage ditches on the landward side of the levee. The excavation of trenches had essentially undermined the landward stability of the levee.

Irrigation district's expert (the author). On the basis of the results of my investigation, the cause of the failure was progressive slope failures (sloughing) of the landward side of the levee. A destabilizing factor was the seepage of water through the levee. The basal slip surface probably developed in a weak layer located near the base of the levee. For example, Fig. 3.16 shows that there is a peat layer (Pt) located about 14 to 15 ft below the top of the levee. Note that the peat layer has a moisture content of 102 percent (Fig. 3.16). This peat layer would have a very low shear strength and could be the basal slip sur-

Figure 3.17   Slope failure of repaired levee (arrow indicates direction of slope movement).

face for the initial sloughing of the levee. It was observed in 1997 that, at the location of the initial levee break, the repaired levee is again experiencing a slope failure of the landward side of the levee as shown in Fig. 3.17. It was also determined that the initial sloughing occurred in the area of a collection pond and that no trenches had been excavated that could have undermined the levee.

Summary.   For this case study of a levee break, the forensic engineers developed three different theories as to the cause of the failure (piping, hydraulic fracturing, and sloughing). This case study also illustrates how a failure theory can affect the proportional responsibility of the different parties in the lawsuit.

3.5   Report Preparation

The written report should summarize the findings of the forensic investigation. Specific sections in the written report may include the following:

- Purpose of the investigation
- Scope of services
- Findings from the document search

- Summary of the site investigation including the observations of damage, data from the nondestructive and destructive testing, and the laboratory test results

- Engineering analyses, including the results of calculations or computer analyses

- Conclusions with a discussion of the cause of failure

- Repair recommendations (if needed)

- Appendixes, which can contain references, test reports by subcontractors, field notes and sketches, laboratory test data, and detailed calculations or computer printouts

Forensic investigation reports are often prepared for clients who do not have an engineering background. In these cases, the report should be simply worded without resorting to engineering jargon. Graphics are especially important for such clients because they may make it easier to visualize the cause of failure. For example, Figs. 3.18 and 3.19 show graphics that explain the cause of damage due to the flooding of Runway 7-25. For this project, the cause of failure was an upward flow of groundwater during a flood that resulted in the development of voids below the asphalt runway. At the time of trial, report graphics like these can be enlarged and presented to a jury so that they can more easily grasp the cause of failure and type of damage.

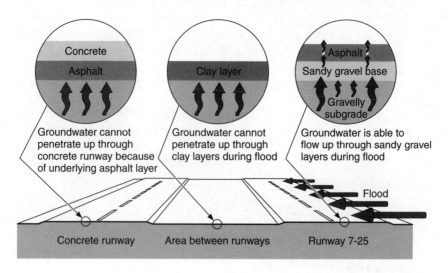

NOT TO SCALE

Figure 3.18 Airport flood condition.

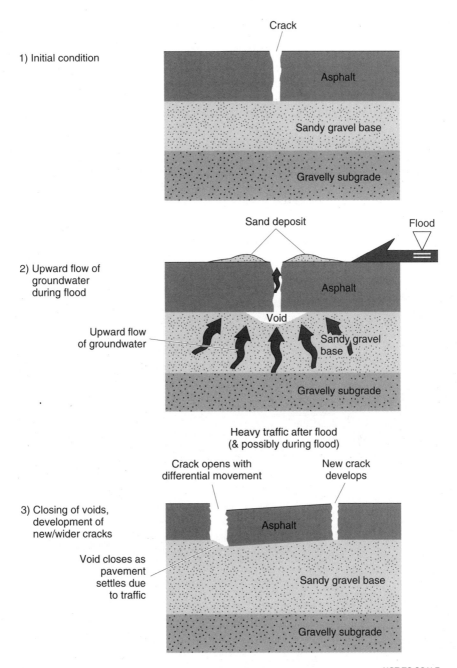

Figure 3.19 Illustration of the cause of damage.

Especially for projects dealing with civil litigation, the report has to be clear, concise, and errorfree. Noon (1992) describes how a report can be used to discredit the forensic engineer at the time of trial:

> The attorney may find a number of small typos, minor errors of fact, etc., in the engineer's report or testimony. He may then argue that the work product is sloppy and error prone, and will then rhetorically ask what other errors or mistakes are imbedded in the engineer's evidence and testimony. This technique is especially effective if the attorney gets the engineer to admit to one important error but then affirm that there are no others. Of course, a prepared attorney will have several more such errors identified and will wait with bated breath to show the engineer those errors, dragging the process out to really make the engineer look careless.

In summary, the report must accurately reflect the findings of the forensic investigation, explain the cause of failure without resorting to engineering jargon, and be errorfree in order to prevent the forensic engineer from being discredited at the time of trial.

Forensic Geotechnical and Foundation Investigations

Slope failure during construction at Oceanside, California. Note that the backpack located in the middle of the photograph provides a scale for the size of the ground cracks.

Settlement of Structures

The following notation is introduced in this chapter:

Symbol	Definition
%C	Percent collapse
e_0	Natural or initial void ratio
Δe_c	Change in void ratio upon wetting
H	Thickness of fill causing settlement
ΔH_c	Change in height upon wetting
H_0	Initial thickness of soil specimen
L	Horizontal distance for calculation of maximum angular distortion
LL	Liquid limit
N	Vertical pressure for consolidation test
PI	Plasticity index
R.C.	Relative compaction
S	Settlement
ε_c	Vertical strain caused by the overburden pressure
ε_s	Vertical strain caused by sampling and laboratory testing
ε_u	Vertical strain due to unloading during sampling
Δ	Maximum differential movement of the foundation
δ	Vertical displacement for calculation of maximum angular distortion
δ/L	Maximum angular distortion of the foundation
γ_d	Dry density
γ_{max}	Laboratory maximum dry density
ρ_{max}	Maximum settlement of the foundation

4.1   Introduction

In terms of forensic engineering, *settlement* is defined as vertical or differential movement of the failed facility resulting in distress or collapse (Greenspan et al., 1989). The determination of settlement can be made through field and laboratory testing. In investigating settlement of structures, it is important to compare the actual applied loading that caused failure with the design or expected loading. Settlement of the structure could be due to increased or unanticipated loading or problems with the bearing soil or rock.

The majority of foundations are adequately constructed and perform as designed; however, there are many instances where settlement can cause damage and foundation failure. Common causes of settlement are consolidation of soft and/or organic soil, settlement from uncontrolled or deep fill, and the development of limestone cavities or sinkholes (Greenfield and Shen, 1992). Foundations can also experience settlement due to natural disasters, such as earthquakes or the undermining of the foundation from floods. There have also been reports of widespread ground subsidence caused by the extraction of oil or groundwater as well as the collapse of underground mines and tunnels.

When investigating damage due to settlement, the forensic engineer should also evaluate the foundation design and construction process, which could be contributing factors in the damage. There are many excellent references, such as *Foundation Analysis and Design* (Bowles, 1982), that present the methods and procedures for the analysis and design of foundations.

4.2   Allowable Settlement

A major reference for the settlement of structures is the paper by Skempton and MacDonald (1956) titled "The Allowable Settlement of Buildings." As shown in Fig. 4.1, Skempton and MacDonald defined the maximum angular distortion δ/L and the maximum differential settlement Δ for a building with no tilt. The *angular distortion* δ/L is defined as the differential settlement between two points divided by the distance between them less the tilt, where tilt equals rotation of the entire building. As shown in Fig. 4.1, the maximum angular distortion δ/L does not necessarily occur at the location of maximum differential settlement (Δ).

Skempton and MacDonald studied 98 buildings, where 58 had suffered no damage and 40 had been damaged in varying degrees as a consequence of settlement. From a study of these 98 buildings, Skempton and MacDonald in part concluded the following:

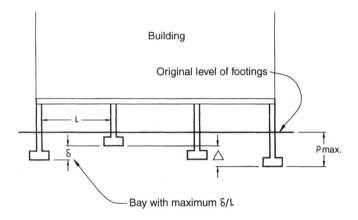

Drawing not to scale

Figure 4.1 Diagram illustrating the definitions of maximum angular distortion and maximum differential settlement.

- The cracking of the brick panels in frame buildings or load-bearing brick walls is likely to occur if the angular distortion of the foundation exceeds 1/300. Structural damage to columns and beams is likely to occur if the angular distortion of the foundation exceeds 1/150.

- By plotting the maximum angular distortion δ/L versus the maximum differential settlement Δ such as shown in Fig. 4.2, a correlation was obtained that is defined as $\Delta = 350\ \delta/L$. (*Note:* Δ is in inches.) On the basis of this relationship and an angular distortion δ/L of 1/300, cracking of brick panels in frame buildings or load-bearing brick walls is likely to occur if the maximum differential settlement Δ exceeds 32 mm ($1\frac{1}{4}$ in.).

- The angular distortion criteria of 1/150 and 1/300 were derived from an observational study of buildings of load-bearing-wall construction, and steel- and reinforced-concrete-frame buildings with conventional brick panel walls but without diagonal bracing. The criteria are intended as no more than a guide for day-to-day work in designing typical foundations for such buildings. In certain cases they may be overruled by visual or other considerations.

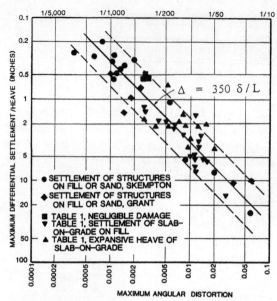

Figure 4.2 Maximum differential settlement versus maximum angular distortion [initial data from Skempton and MacDonald (1956), Table 1 in Day (1990a)].

The paper by Grant et al. (1974) updated the Skempton and MacDonald data pool and also evaluated the rate of settlement with respect to the amount of damage incurred. Grant et al. (1974) in part concluded the following:

- A building foundation that experiences a maximum value of deflection slope δ/L greater than 1/300 will probably suffer some damage. However, damage does not necessarily occur at the point where the local deflection slope exceeds 1/300.

- For any type of foundation on sand or fill, new data tend to support Skempton and MacDonald's suggested correlation of $\Delta = 350\ \delta/L$ (see Fig. 4.2).

- Consideration of the rate of settlement is important only for the extreme situations of either very slow or very rapid settlement. On the basis of the limited data available, the values of maximum δ/L corresponding to building damage appear to be essentially the same for cases involving slow and fast settlements.

Data concerning the behavior of lightly reinforced, conventional slab-on-grade foundations have also been included in Fig. 4.2. This data indicates that cracking of gypsum wallboard panels is likely to occur if the angular distortion of the slab-on-grade foundation exceeds 1/300 (Day, 1990a). The ratio of 1/300 appears to be useful for both wood-frame gypsum wallboard panels and the brick panels studied by Skempton and MacDonald (1956). The data plotted in Fig. 4.2 would indicate that the relationship $\Delta = 350\ \delta/L$ can also be used for buildings supported by lightly reinforced slab-on-grade foundations. Using $\delta/L = 1/300$ as the boundary where cracking of panels in wood-frame residences supported by concrete slab-on-grade is likely to occur and substituting this value into the relationship $\Delta = 350\ \delta/L$ (Fig. 4.2), we calculate a differential slab displacement of 32 mm ($1\frac{1}{4}$ in.). For buildings on lightly reinforced slabs-on-grade, cracking of gypsum wallboard panels is likely to occur when the maximum slab differential exceeds 32 mm ($1\frac{1}{4}$ in.). As shown in Fig. 4.3, this cracking frequently develops at the locations of stress concentrations, such as at the corners of door or window openings.

4.2.1 Different types of foundations and buildings

There is other data available on the allowable settlement of structures (e.g., Leonards, 1962; ASCE, 1964; Feld, 1965; Peck et al., 1974; Wahls, 1994). For example, it has been stated that the allowable differential and total settlement should depend on the flexibility and

Figure 4.3 Interior wallboard cracking.

complexity of the structure including the construction materials and type of connections (*Foundation Engineering Handbook*, Winterkorn and Fang, 1975). In this regard, Terzaghi (1938) states:

> Differential settlement must be considered inevitable for every foundation, unless the foundation is supported by solid rock. The effect of the differential settlement on the building depends to a large extent on the type of construction.

Terzaghi (1938) summarized his studies on several buildings in Europe where he found that walls 18 m (60 ft) and 23 m (75 ft) long with differential settlements over 2.5 cm (1 in.) were all cracked, but four buildings with walls 12 m (40 ft) to 30 m (100 ft) long were undamaged when the differential settlement was 2 cm ($^3/_4$ in.) or less. This is probably the basis for the general design guide that building foundations should be designed so that the differential settlement is 2 cm ($^3/_4$ in.) or less.

Another example of allowable settlements for buildings is Table 4.1 (from Sowers, 1962). In this table, the allowable foundation displacement has been divided into three categories: total settlement, tilting, and differential movement. Table 4.1 indicates that those structures that are more flexible (such as simple steel frame buildings) or have more rigid foundations (such as mat foundations) can sustain larger values of total settlement and differential movement.

Figure 4.4 presents data from Bjerrum (1963). Like the studies previously mentioned, this figure indicates that cracking in panel walls is to be expected at an angular distortion δ/L of 1/300 and that structural damage of buildings is to be expected at an δ/L of 1/150. This figure also provides other limiting values of δ/L, such as for buildings containing sensitive machinery or overhead cranes.

4.3   Classification of Cracking Damage

Table 4.2 summarizes the severity of cracking damage versus approximate crack widths, typical values of maximum differential movement Δ, and maximum angular distortion δ/L of the foundation (Burland et al., 1977; Boone, 1996; Day, 1998a). The relationship between differential settlement Δ and angular distortion δ/L was based on the equation $\Delta = 350\ \delta/L$ (from Fig. 4.2). There may be cases where the damage category will not correlate with Δ or δ/L. Common examples include those where the cracking has been hidden or patched and where other factors (such as concrete shrinkage) contribute to crack widths.

TABLE 4.1   Allowable Settlement

Type of movement (1)	Limiting factor (2)	Maximum settlement (3)
Total settlement	Drainage	15–30 cm (6–12 in.)
	Access	30–60 cm (12–24 in.)
	Probability of nonuniform settlement:	
	Masonry-walled structure	2.5–5 cm (1–2 in.)
	Framed structures	5–10 cm (2–4 in.)
	Smokestacks, silos, mats	8–30 cm (3–12 in.)
Tilting	Stability against overturning	Depends on H and W
	Tilting of smokestacks, towers	$0.004L$
	Rolling of trucks, etc.	$0.01L$
	Stacking of goods	$0.01L$
	Machine operation—cotton loom	$0.003L$
	Machine operation—turbogenerator	$0.0002L$
	Crane rails	$0.003L$
	Drainage of floors	$0.01–0.02L$
Differential movement	High continuous brick walls	$0.0005–0.001L$
	One-story brick mill building, wall cracking	$0.001–0.002L$
	Plaster cracking (gypsum)	$0.001L$
	Reinforced-concrete building frame	$0.0025–0.004L$
	Reinforced-concrete building curtain walls	$0.003L$
	Steel frame, continuous	$0.002L$
	Simple steel frame	$0.005L$

NOTES: L = distance between adjacent columns that settle different amounts, or between any two points that settle differently. Higher values are for regular settlements and more tolerant structures. Lower values are for irregular settlement and critical structures. H = height and W = width of structure.
SOURCE: Sowers, 1962.

As previously mentioned in Sec. 1.2, there are three general categories of visible damage: architectural, functional (or serviceability), and structural. In Table 4.2, architectural damage is commonly associated with a damage category of at least "very slight." Functional damage is commonly associated with a damage category of at least "slight." Structural damage can be present when the damage category is "moderate," but is most common in the "severe" to "very severe" damage categories.

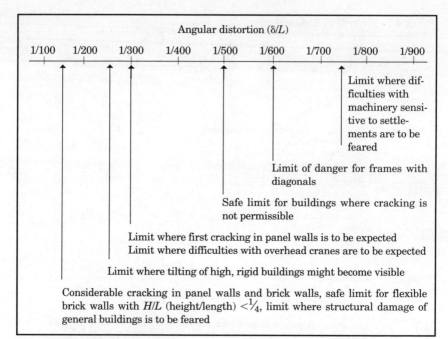

Figure 4.4 Damage criteria. (*After Bjerrum, 1963.*)

4.4 Component of Lateral Movement

Foundations subjected to settlement are usually damaged by a combination of both vertical and horizontal movements. For example, a common cause of foundation damage is fill settlement. Figure 4.5 shows an illustration of the settlement of fill in a canyon environment. Over the sidewalls of the canyon, there tends to be a pulling or stretching of the ground surface (tensional features), with compression effects near the canyon centerline. This type of damage is due to two-dimensional settlement, where the fill compresses in both the vertical and horizontal directions (Lawton et al., 1991; Day, 1991a). Figures 4.6 and 4.7 show typical foundation cracking and structural damage to a building experiencing fill settlement. Note in Figs. 4.6 and 4.7 that the fill settlement causes both vertical and horizontal displacements, which tend to pull apart and distort the building structure.

Another common situation where both vertical and horizontal foundation displacement occurs is at cut-fill transitions. A cut-fill transition occurs when a building pad has some rock removed (the cut portion), with a level building pad being created by filling in (with soil)

TABLE 4.2   Severity of Cracking Damage

Damage category (1)	Description of typical damage (2)	Approx. crack width (3)	Δ (4)	δ/L (5)
Negligible	Hairline cracks	<0.1 mm	<3 cm (1.2 in.)	<1/300
Very slight	Very slight damage includes fine cracks that can be easily treated during normal decoration, perhaps an isolated slight fracture in building, and cracks in external brickwork visible on close inspection.	1 mm	3 to 4 cm (1.2–1.5 in.)	1/300 to 1/240
Slight	Slight damage includes cracks that can be easily filled and redecoration would probably be required; several slight fractures may appear showing on the inside of the building; cracks that are visible externally and some repointing may be required; and doors and windows may stick.	3 mm	4 to 5 cm (1.5–2.0 in.)	1/240 to 1/175
Moderate	Moderate damage includes cracks that require some opening up and can be patched by a mason; recurrent cracks that can be masked by suitable linings; repointing of external brickwork and possibly a small amount of brickwork replacement may be required; doors and windows stick; service pipes may fracture; and weathertightness is often impaired.	5 to 15 mm or a number of cracks >3 mm	5 to 8 cm (2.0–3.0 in.)	1/175 to 1/120
Severe	Severe damage includes large cracks requiring extensive repair work involving breaking out and replacing sections of walls (especially over doors and windows); distorted windows and door frames; noticeably sloping floors; leaning or bulging walls; some loss of bearing in beams; and disrupted service pipes.	15 to 25 mm but also depends on number of cracks	8 to 13 cm (3.0–5.0 in.)	1/120 to 1/70
Very severe	Very severe damage often requires a major repair job involving partial or complete rebuilding; beams lose bearing; walls lean and require shoring; windows are broken with distortion; and there is danger of structural instability.	Usually >25 mm but also depends on number of cracks	>13 cm (5 in.)	>1/70

the remaining portion. If the cut side of the building pad contains nonexpansive rock that is dense and unweathered, then very little settlement would be expected for that part of the building on cut. But the fill portion could settle under its own weight and cause damage. For example, a slab crack will typically open at the location of the cut-fill transition as illustrated in Fig. 4.8. The building is damaged by both the vertical foundation movement (settlement) and the horizon-

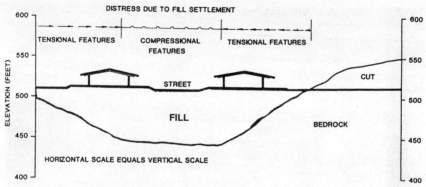

Figure 4.5 Fill settlement in a canyon environment.

Figure 4.6 Fill settlement: foundation damage.

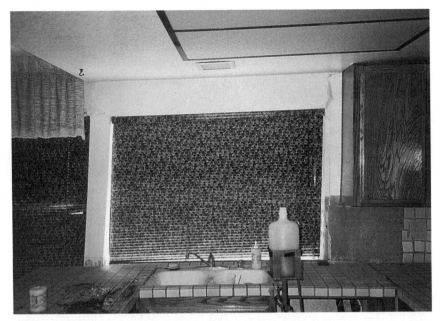

Figure 4.7 Fill settlement: structural damage.

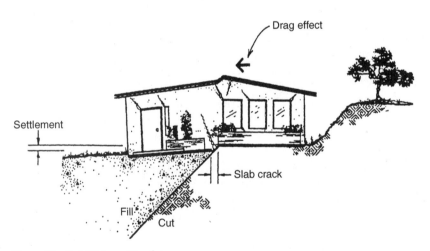

Figure 4.8 Cut-fill transition lot.

tal movement, which manifests itself as a slab crack-and-drag effect on the structure (Fig. 4.8).

In the cases described above, the lateral movement is a secondary result of the primary vertical movement due to settlement of the foundation. Table 4.2 can therefore be used as a guide to correlate damage category with Δ and δ/L.

4.5 Collapsible Soil

In the southwestern United States, probably the most common cause of settlement is collapsible soil. For example, Johnpeer (1986) states that ground subsidence from collapsing soils is a common occurrence in New Mexico. The category of collapsible soil would include the settlement of debris, uncontrolled fill, deep fill, or natural soil, such as alluvium or colluvium.

In general, there has been an increase in damage due to collapsible soil, probably because of the lack of available land in many urban areas. This causes development of marginal land, which may contain deposits of dumped fill or deposits of natural collapsible soil. Also, substantial grading can be required to develop level building pads, which results in more areas having deep fill.

Collapsible soil can be broadly classified as soil that is susceptible to a large and sudden reduction in volume upon wetting. Collapsible soil usually has a low dry density and low moisture content. Such soil can withstand a large applied vertical stress with a small compression, but then experience much larger settlements after wetting, with no increase in vertical pressure (Jennings and Knight, 1957).

Fill. Deep fill has been defined as fill that has a thickness greater than 6 m (20 ft) (Greenfield and Shen, 1992). Uncontrolled fills include fills that were not documented with compaction testing as they were placed; these include dumped fills, fills dumped under water, hydraulically placed fills, and fills that may have been compacted but there is no documentation of testing or the amount of effort that was used to perform the compaction (Greenfield and Shen, 1992). These conditions may exist in rural areas where inspections are lax or for structures built many years ago when the standards for fill compaction were less rigorous.

For collapsible fill, compression will occur as the overburden pressure increases. The increase in overburden pressure could be due to the placement of overlying fill or the construction of a building on top of the fill. The compression due to this increase in overburden pressure involves a decrease in void ratio of the fill due to expulsion of air.

The compression usually occurs at constant moisture content. After completion of the fill mass, water may infiltrate the soil due to irrigation, rainfall, or leaky water pipes. The mechanism that usually causes the collapse of the loose soil structure is a decrease in negative pore pressure (capillary tension) as the fill becomes wet.

For a fill specimen submerged in distilled water, the main variables that govern the amount of one-dimensional collapse are the soil type, compacted moisture content, compacted dry density, and the vertical pressure (Dudley, 1970; Lawton et al., 1989, 1991, 1992; Tadepalli et al., 1991; Day, 1994a). In general, the one-dimensional collapse of fill will increase as the dry density decreases, the moisture content decreases, or the vertical pressure increases. For a constant dry density and moisture content, the one-dimensional collapse will decrease as the clay fraction increases once the optimum clay content (usually a low percentage) is exceeded (Rollins et al., 1994).

Alluvium or colluvium.   For natural deposits of collapsible soil in the arid climate of the Southwest, a common mechanism involved in rapid volume reduction entails breaking of bonds at coarse particle contacts by weakening of fine-grained materials brought there by surface tension in evaporating water. In other cases, the alluvium or colluvium may have an unstable soil structure which collapses as the wetting front passes through the soil.

Laboratory testing.   If collapsible soil is the suspected cause of damage at a site, then soil specimens should be obtained and tested in the laboratory. One-dimensional collapse is usually measured in the oedometer (ASTM D 5333-92, "Standard Test Method for Measurement of Collapse Potential of Soils," 1997). After the soil specimen is placed in the oedometer, the vertical pressure is increased until it approximately equals the *in situ* overburden pressure. At this vertical pressure, distilled water is added to the oedometer to measure the amount of collapse of the soil specimen. Percent collapse ($\%C$) is defined as the change in height of the specimen due to inundation divided by the initial height of the specimen (ASTM D 5333-92, 1997).

Figure 4.9 presents the results of a one-dimensional collapse test performed on a fill specimen. The fill specimen contains 60 percent sand size particles, 30 percent silt size particles, and 10 percent clay size particles and is classified as a silty sand (SM). To model field conditions, the silty sand was compacted at a dry density of 1.48 Mg/m^3 (92.4 pcf) and moisture content of 14.8 percent. The silty sand specimen, having an initial height of 25.4 mm (1.0 in.), was subjected to a vertical stress of 144 kPa (3000 psf) and then inundated with distilled

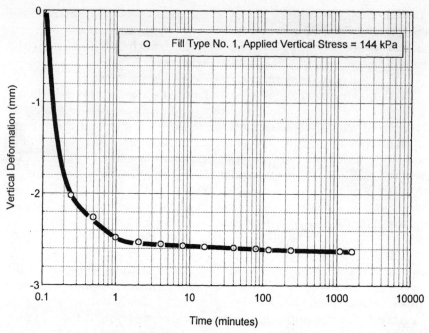

Figure 4.9 Typical collapse potential test results.

water. Figure 4.9 shows the amount of vertical deformation (collapse) as a function of time after inundation. Collapse is calculated as Δe_c divided by $1 + e_0$, or as ΔH_c divided by H_0, expressed as a percentage. For the collapse test on the silty sand (fill type number 1, Fig. 4.9), $(\%C) = (2.62 \text{ mm})/(25.4 \text{ mm}) = 10.3\%$.

The collapse potential (CP) of a soil can be determined by applying a vertical stress of 200 kPa (2 tsf) to the soil specimen, submerging it in distilled water, and then determining the percent collapse which is designated the collapse potential. This collapse potential can be considered as an index test to compare the susceptibility of collapse for different soils. A CP of 4 to 6% indicates moderate collapse potential, while a CP over 10% indicates severe collapse potential.

As previously mentioned, the triggering mechanism for the collapse of fill or natural soil is the introduction of moisture. Common causes of the infiltration of moisture include water from irrigation, broken or leaky water lines, and an altering of surface drainage which allows rainwater to pond near the foundation. Another source of moisture infiltration is leaking pools. For example, Fig. 4.10 shows a photograph of a severely damaged pool shell which cracked due to the collapse of an underlying uncompacted debris fill.

Figure 4.10 Damage to a pool shell due to collapse of underlying uncompacted fill (arrows point to cracks in the pool shell).

4.6 Backfill Settlement

Heavy compaction equipment is commonly used to compact fill for mass grading projects (Monahan, 1986). But many projects require backfill compaction where the construction space is too small to allow for such heavy equipment. Some examples include the backfill for utility trenches excavated in roads or the backfill behind retaining walls. For these projects, the backfill may be simply dumped in place or compacted with minimal compaction energy by hand tampers. In addition, the backfill is commonly not tested to evaluate the quality of the compaction. These factors of limited access, poor compaction process, and lack of compaction testing frequently lead to backfill settlement. As an example of backfill settlement, Fig. 4.11 shows a sketch where gravel was used as backfill material behind a garage wall. At this site, the gravel was simply dumped behind the garage wall and then stairs were constructed atop the gravel backfill. Vibrations due to subsequent construction and the migration of water down through the gravel caused the loose gravel to settle resulting in the damage shown in Figs. 4.12 and 4.13. The settlement of the gravel backfill could have been avoided if the gravel had been placed in layers and then compacted by a vibrating hand tamper.

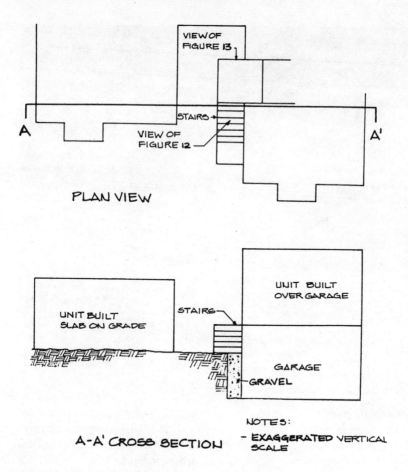

Figure 4.11   Sketch showing location of gravel backfill.

4.7   Other Causes of Settlement

Besides collapsible soil or backfill settlement, there can be many other causes of settlement that damage structures. There are many excellent references to help the forensic engineer identify a particular cause of settlement. Some of the more common causes of settlement are discussed below.

4.7.1   Limestone cavities or sinkholes

Settlement related to limestone cavities or sinkholes will usually be limited to areas having karst topography. Karst topography is a type of landform developed in a region of easily soluble limestone bedrock.

Figure 4.12 Damage to stairs caused by backfill settlement.

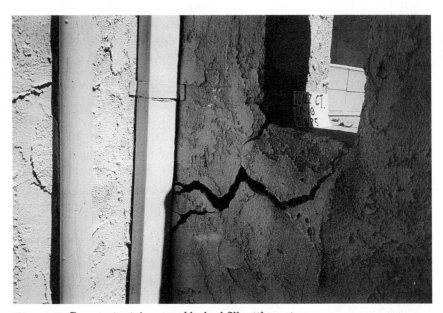

Figure 4.13 Damage to stairs caused by backfill settlement.

It is characterized by vast numbers of depressions of all sizes, sometimes by great outcrops of limestone, sinks and other solution passages, an almost total lack of surface streams, and larger springs in the deeper valleys (Stokes and Varnes, 1955).

Identification techniques and foundations constructed on karst topography are discussed by Sowers (1997). Methods to investigate the presence of sinkholes are geophysical techniques and the cone penetration device (*Earth Manual,* 1985; Foshee and Bixler, 1994). A low cone penetration resistance could indicate the presence of raveling, which is a slow process where granular soil particles migrate into the underlying porous limestone. An advanced state of raveling will result in subsidence of the ground, commonly referred to as *sinkhole activity.*

4.7.2 Consolidation of soft and/or organic soil

Most soil mechanics textbooks have an in-depth discussion of the subsurface exploration, laboratory testing, and engineering analysis required to identify and evaluate the settlement potential of soft clay and/or organic soil (Terzaghi and Peck, 1967; Lambe and Whitman, 1969; Sowers, 1979; Holtz and Kovacs, 1981; Cernica, 1995a). The settlement of saturated clay or organic soil can have three different components: immediate (or initial), consolidation, and secondary compression.

Immediate or initial. In most situations, surface loading causes both vertical and horizontal strains, and this is referred to as *two-* or *three-dimensional loading.* Immediate settlement is due to undrained shear deformations, or in some cases contained plastic flow, caused by the two- or three-dimensional loading (Ladd et al., 1977). Common examples of two-dimensional loading are from square footings and round storage tanks. Such a loading of a saturated clay can result in distress or collapse due to vertical and horizontal strains caused by immediate settlement of the clay.

Consolidation. The typical one-dimensional case of settlement involves strain in only the vertical direction. Common examples of one-dimensional loading include the lowering of the water table or a uniform fill surcharge applied over a very large area. Consolidation is a time-dependent process that may take many years to complete. For example, Fig. 4.14 shows the vertical deformation versus time from a consolidation test (ASTM D 2435-96, 1997) performed on a saturated clay (PI = 71, LL = 93, H_0 = 9.0 mm, e_0 = 3.05).

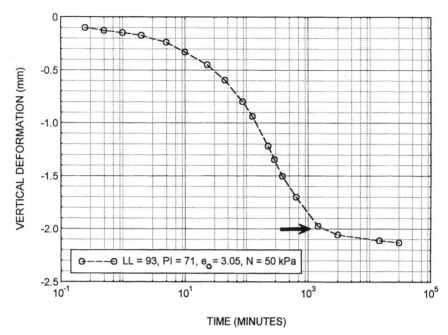

TIME (MINUTES)

Figure 4.14 Typical S-type deformation versus time curve (arrow indicates end of primary consolidation).

Secondary compression. The final component of settlement is due to secondary compression, which is that part of the settlement that occurs after essentially all of the excess pore water pressures have dissipated. Secondary compression occurs at constant effective stress and can constitute a major part of the total settlement for peat or other highly organic soil (Holtz and Kovacs, 1981). Figure 4.15 shows two examples of the settlement of peat at the Meadowlands, a marshy area in New Jersey (west of New York City). Piles are often used to support structures at the Meadowlands, but as shown in Fig. 4.15, the floor slab is typically unable to span between piles and it breaks off from the pile caps or it deforms around the pile caps as the peat settles (Whitlock and Moosa, 1996).

4.7.3 Collapse of underground mines and tunnels

According to Gray (1988), damage to residential structures in the United States caused by the collapse of underground mines is estimated to be between \$25 and \$35 million each year, with another \$3 to \$4 million in damage to roads, utilities, and services. There are

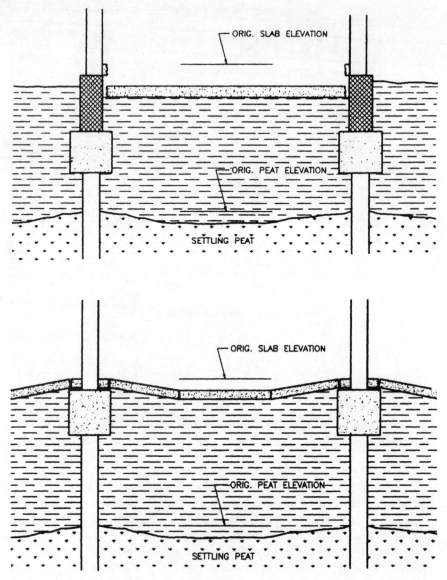

Figure 4.15 Common slab displacement mechanisms due to subsoil subsidence. [*From Whitlock and Moosa (1996), reprinted with permission from the American Society of Civil Engineers.*]

approximately 2 million hectares of abandoned or inactive coal mines, with 10 percent of these hectares in populated urban areas (Dyni and Burnett, 1993). It has been stated that ground subsidence associated with longwall mining can be predicted fairly well with respect to magnitude, time, and a real position (Lin et al., 1995). Once the amount of ground subsidence has been estimated, there are measures that can be taken to mitigate the effects of mine-related subsidence (National Coal Board, 1975; Kratzsch, 1983; Peng, 1986, 1992). For example, in a study of different foundations subjected to mining-induced subsidence, it was concluded that posttensioning of the foundation was most effective, because it prevented the footings from cracking (Lin et al., 1995).

As in the discussion presented in Sec. 4.4, the collapse of underground mines and tunnels can produce tension and compression-type features within the buildings. Figure 4.16 (from Marino et al., 1988) shows that the location of the compression zone will be in the center of the subsided area. The tension zone is located along the perimeter of the subsided area. These tension and compression zones are similar to fill settlement in a canyon environment (Fig. 4.5).

Besides the collapse of underground mines and tunnels, there can be settlement of buildings constructed on spoil extracted from the mines. Mine operators often dispose of other debris, such as trees, scrap metal, and tires, within the mine spoil. In many cases, the mine spoil is dumped (no compaction) and can be susceptible to large

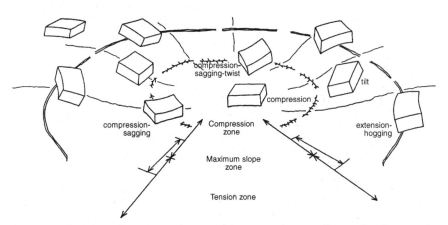

Figure 4.16 Location of tension and compression zones due to collapse of underground mines. [*From Marino et al. (1988), reprinted with permission from the American Society of Civil Engineers.*]

amounts of settlement. For example, Cheeks (1996) describes an interesting case of a motel unknowingly built on spoil that had been used to fill in a strip-mining operation. The motel building experienced about 1 m (3 ft) of settlement within the monitoring period (5 years). The settlement and damage for this building actually started during construction and the motel owners could never place the building into service. The outcome of the subsequent lawsuit was that the cost of litigation exceeded $1 million which approached the value of the final monetary award (Cheeks, 1996). Litigation expenses that approach or even exceed the final judgment are a common occurrence.

4.7.4 Ground subsidence due to extraction of oil or groundwater

Large-scale pumping of water or oil from the ground can cause settlement of the ground surface over a large area. The pumping can cause a lowering of the groundwater table which then increases the overburden pressure on underlying sediments. This can cause consolidation of soft clay deposits. In other cases, the removal of water or oil can lead to compression of the soil or porous rock structure, which then results in ground subsidence.

Lambe and Whitman (1969) describe two famous cases of ground surface subsidence due to oil or groundwater extraction. The first is oil pumping from Long Beach, California, which affected a 65 km^2 (25 mi^2) area and caused 8 m (25 ft) of ground surface subsidence. Because of this ground surface subsidence, the Long Beach Naval Shipyard had to construct special sea walls to keep the ocean from flooding the facilities. A second famous example is ground surface subsidence caused by pumping of water for domestic and industrial use in Mexico City. Rutledge (1944) shows that the underlying Mexico City clay, which contains a porous structure of microfossils and diatoms, has a very high void ratio (up to $e_0 = 14$) and is very compressible. Ground surface subsidence in Mexico City has been reported to be 9 m (30 ft) since the beginning of the twentieth century.

Besides ground surface subsidence, the extraction of groundwater or oil can also cause the opening of ground fissures. For example, Fig. 4.17 shows the ground surface subsidence in Las Vegas Valley between 1963 and 1987 due primarily to groundwater extraction. It has been stated (Purkey et al., 1994) that the subsidence has been focused on preexisting geologic faults, which serve as points of weakness for ground movement. Figure 4.18 shows one of these fissures that ran beneath the foundation of a home.

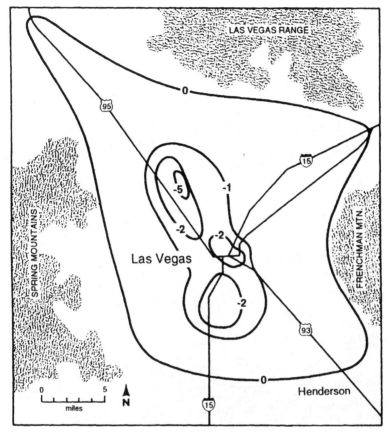

Figure 4.17 Land surface subsidence (in feet) in Las Vegas Valley between 1963 and 1987. [*From Purkey et al. (1994), reprinted with permission from the Nevada Bureau of Mines and Geology.*]

4.7.5 Settlement of landfills and decomposition of organic matter

As previously mentioned, one common cause of settlement is consolidation and secondary compression of soft and saturated organic soils. This type of settlement is a result of the drainage of water from the void spaces due to an increase in overburden pressure. This is a different mechanism than settlement caused by the decomposition of organic matter.

Decomposition of organic matter. Organic matter consists of a mixture of plant and animal products in various stages of decomposition, of

Figure 4.18 Fissure, widened by erosion, running beneath the foundation of an abandoned home near Simmons St. in North Las Vegas. [*From Purkey et al. (1994), reprinted with permission from the Nevada Bureau of Mines and Geology.*]

substances formed biologically and/or chemically from the breakdown of products, and of microorganisms and small animals and their decaying remains. For this very complex system, organic matter is generally divided into two groups: nonhumic and humic substances (Schnitzer and Khan, 1972). Nonhumic substances include numerous compounds such as carbohydrates, proteins, and fats that are easily attacked by microorganisms in the soil and normally have a short existence. Most organic matter in soils consists of humic substances, defined as amorphous, hydrophilic, acidic, and polydisperse substances (Schnitzer and Khan, 1972). An example of humic substances is the brown or black part of organic matter, which is so well decomposed that the original source cannot be identified. The important characteristics exhibited by humic substances are a resistance to microbic deterioration, and the ability to form stable compounds (Kononova, 1966).

The forensic engineer can identify organic matter by its brown or black color, pungent odor, spongy feel, and in some cases its fibrous texture. The change in character of organic matter due to decomposition has been studied by Al-Khafaji and Andersland (1981). They present visual evidence of the changes in pulp fibers due to microbic activity, using a scanning electron microscope (SEM). Decomposition of pulp fibers includes a reduction in length, diameter, and the development of rougher surface features. The ignition test is commonly used to determine the percent organics (ASTM D 2974-95, 1997). In this test the humic and nonhumic substances are destroyed by high ignition temperatures. The main source of error in the ignition test is the loss of surface hydration water from the clay minerals. Franklin et al., (1973) indicate that large errors can be produced if certain minerals, such as montmorillonite, are present in quantity. The ignition test also does not distinguish between the humic and nonhumic fraction of organics.

The problem with decomposing organics is the development of voids and corresponding settlement. The rate of settlement will depend on how fast the nonhumic substances decompose, and the compression characteristics of the organics. For large-mass graded projects it is difficult to keep all organic matter out of the fill. The detrimental effects of organic matter are generally recognized by most engineers. Common organic matter inadvertently placed in fill includes branches, shrubs, leaves, grass, and construction debris such as pieces of wood and paper. Figure 4.19 shows a photograph of decomposing organics placed in a structural fill.

Landfills. Landfills can settle because of compression of the loose waste products and decomposition of organic matter that was placed within the landfill. The study of landfills provides data on the rate of decomposition of large volumes of organic matter. An excavation at the Mallard North Landfill in Hanover Park, Illinois, unearthed a 15-year-old steak (dated by a legible newspaper buried nearby) that "had bone, fat, meat, everything" (Rathje and Psihoyos, 1991). The lack of decomposition of these nonhumic substances can be attributed to the fact that a typical landfill admits no light and little air or moisture, so the organic matter in the trash decomposes slowly. The nonhumic substances will decompose eventually, but because little air (and

Figure 4.19 Decomposing organic matter placed in a structural fill.

hence oxygen) circulates around the organics, the slower-working anaerobic microorganisms thrive.

Rathje and Psihoyos (1991) have found that 20 to 50 percent of food and yard waste biodegrades in the first 15 years. There can be exceptions such as the Fresh Kills landfill, opened by New York City in 1948, on Staten Island's tidal marshland. Below a certain level in the Fresh Kills landfill, there is no food debris or yard wastes, and practically no paper. The reason is, apparently, that water from the tidal wetlands has seeped into the landfill, causing the anaerobic microorganisms to flourish. The study by Rathje and Psihoyos (1991) indicates that the type of environment is very important in terms of how fast the nonhumic substances decompose.

There are many old abandoned landfills throughout the United States. Typical forensic cases involve the construction of buildings on such abandoned landfills. Settlement of structures constructed atop municipal landfill could be due to compression of the underlying loose waste products or the decomposition of any organic matter remaining in the landfill.

4.8 Change in Properties with Time

Besides determining the cause of distress or collapse of the structure, the forensic engineer may also be asked to determine possible negligence on the part of the design engineer. Many soil properties or construction standards can change with time. It is important for the forensic engineer to apply those standards or assess the soil properties that existed at the time of construction.

4.8.1 Example of change in properties with time

The following is an example of how properties can change with time. The example deals with fill compaction. A common recommendation is to compact fill to a specific relative compaction. The definition of relative compaction is the as-compacted field dry density divided by the laboratory maximum dry density, usually expressed as a percentage. In California, typical mass grading specifications require a minimum relative compaction of 90 percent, using the modified Proctor laboratory compaction test. The specifications adopted by city and county agencies are from the *Uniform Building Code* (1997), which states: "All fills shall be compacted to a minimum of 90 percent of maximum density."

There are several possible ways that the relative compaction can change with time:

- *Change of in situ fill density.* Examples of how the *in situ* density can increase are as follows: (1) there could be compression of the fill caused by the weight of overlying fill or structures, (2) there could be settlement of the fill due to collapse, or (3) there could be vibrations or seismic activity that densify granular fill. Fill can also change density during sampling or testing. For example, a loose fill may be densified when a sampling tube is driven into the deposit. Likewise, a dense fill could be loosened during the same sampling process.

- *Change in laboratory maximum dry density standard.* The most common change in the maximum dry density with time results from the adoption of new standards. For example, in many localities, the standard Proctor (ASTM D 698, 1997) has been replaced by the modified Proctor (ASTM D 1557, 1997). Using the wrong specification would result in a different value of laboratory maximum dry density.

- *Oversize particle content.* There can also be a change in the maximum dry density for fill containing oversize particles, defined as particles retained on the 19-mm (3/4-in.) sieve. One problem is that there are several different methods that can be used to calculate the laboratory maximum dry density when a soil has oversize particles. Laboratory tests performed by Houston and Walsh (1993) show a substantial difference in laboratory maximum dry density, depending on the oversize correction method.

■ *Breakdown of oversize particles.*   When certain types of oversize particles, such as weak sedimentary rock, are compacted, there can be significant breakdown of oversize particles due to the mechanical process of compaction as well as the subsequent weathering of the oversize particles (Saxena et al., 1984). For weakly cemented sedimentary rock, the degradation of the oversize material produces a better graded material which, when tested long after initial placement, will result in a higher maximum dry density.

All of these factors will have to be considered by the forensic engineer in determining the appropriate laboratory maximum dry density that is applicable at the time of construction.

The following equation can be used by the forensic engineer to estimate the relative compaction (R.C.) of the fill at the time of placement:

$$\text{R.C.} = \frac{100\,\gamma_d}{[1 + (S/H) + \Delta\varepsilon_v + \varepsilon_s]\,\gamma_{max}} \tag{4.1}$$

where γ_d = measured dry density of the fill determined by the forensic engineer, usually at a time well after the project has been completed

S = measured vertical settlement (positive value) or measured vertical heave (negative value) for a zone of soil movement

H = zone of fill settlement or heave

ε_s = vertical strain caused by sampling and laboratory testing, where a positive value is net compression, while a negative value is net volume increase

γ_{max} = laboratory maximum dry density, taking into account the possible change in standards with time and a possible change in grain size distribution

The value $\Delta\varepsilon_v$ is defined as

$$\Delta\varepsilon_v = \varepsilon_c - \varepsilon_u \tag{4.2}$$

where ε_c = vertical strain due to the overburden pressure and ε_u = vertical strain due to unloading during sampling.

4.9   Case Studies

Slab-on-grade foundation.   The following is a case study of damage to a house caused by settlement, illustrating the use of Eqs. (4.1) and

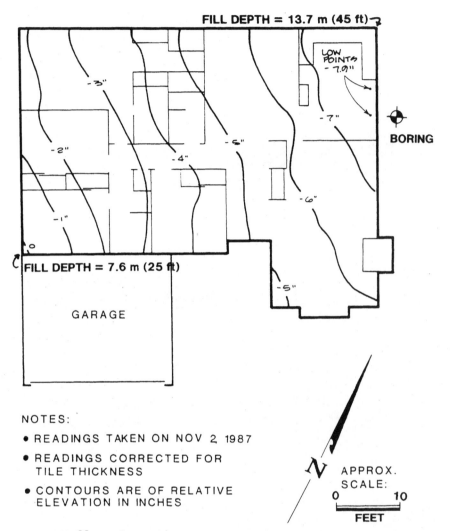

FILL DEPTH = 13.7 m (45 ft)

LOW POINTS -7.9"

-3"

-7"

BORING

-2"

-4"

-6"

-1"

-6"

0

FILL DEPTH = 7.6 m (25 ft)

-5"

GARAGE

NOTES:

• READINGS TAKEN ON NOV 2 1987

• READINGS CORRECTED FOR TILE THICKNESS

• CONTOURS ARE OF RELATIVE ELEVATION IN INCHES

N

APPROX. SCALE:

0   10

FEET

Figure 4.20 Manometer survey.

(4.2). A level building pad was created in 1973 to 1974 by filling in a deep canyon. The house was then constructed on a lightly reinforced, conventional slab-on-grade foundation. The depth of fill underneath the house, based on the original grading plans and subsurface exploration, varied from 7.6 m (25 ft) up to 13.7 m (45 ft). Figure 4.20 shows a sketch of the house and indicates the location of the shallowest and deepest fill.

By 1987, the house had experienced 20 cm (7.9 in.) of differential foundation movement (Δ). This movement caused very severe dam-

Figure 4.21 Cracking of exterior stucco at garage door.

age, such as foundation cracks, interior wallboard cracks, racked door frames, and a near complete separation of the house from the originally attached garage. The fill settlement caused both functional and structural damage. A picture of typical observed damage is shown in Fig. 4.21.

A benchmark had been established by the city on a nearby sidewalk in 1975. This benchmark recorded 32 cm (12.6 in.) of settlement from 1975 to 1987.

A large-diameter boring excavated adjacent the residence (see Fig. 4.20) revealed clayey sand to sandy clay fill underlaid by very dense bedrock. At a depth of 10.5 m (34.5 ft), an undisturbed block sample of the fill was retrieved from the large-diameter boring. The block sample was transported to the laboratory and the dry density of the fill was determined to be 1.61 Mg/m^3 (100.7 pcf). The laboratory maximum dry density was 1.95 Mg/m^3 (121.5 pcf) using the modified Proctor test standard (ASTM D 1557, 1997), which was the project specification at the time of fill placement. The calculated relative compaction would be 1.61/1.95 = 83%.

Equation (4.1) can be used to estimate the relative compaction at the time of fill placement in 1973 to 1974. The following parameters were used:

γ_d = 1.61 Mg/m^3 (100.7 pcf), from laboratory tests on the block sample

S = 23 cm (9 in.), based on the measured differential settlement of the house (Fig. 4.20) and the nearby city-established monument

H = 4.6 m (15 ft), from the subsurface exploration which revealed the zone of poorly compacted fill from a depth of about 7.6 m (25 ft) to 12.2 m (40 ft)

ε_c = 4.5%, from the loading curve of a compression test performed on a specimen trimmed from the block sample

ε_u = 2.0%, from the unloading curve of the previously mentioned compression test

$\Delta\varepsilon_u = \varepsilon_c - \varepsilon_u = 4.5\% - 2\% = 2.5\%$

ε_s = 0%, because an undisturbed block sample was used

γ_{max} = 1.95 Mg/m^3 (121.5 pcf), from the modified Proctor compaction test

Substituting the above values into Eq. (4.1) gives an estimated value of relative compaction at the time of fill placement in 1973 to 1974 of 77 percent. This value is well below the required value of 90 percent. It was concluded for this project that the cause of settlement was the placement of poorly compacted (R.C. = 77%) fill. The relative compaction was so low that it was also concluded that essentially negligible compaction effort was used during construction.

Posttensioned slabs-on-grade. Posttensioned slabs-on-grade are common in southern California and other parts of the United States. They are an economical foundation type when there is no ground freezing or the depth of frost penetration is low. The most common uses of posttensioned slabs-on-grade are to resist expansive soil forces or when the projected differential settlement exceeds the tolerable value for a conventional (lightly reinforced) slab-on-grade. For example, posttensioned slabs-on-grade are frequently recommended if the projected differential settlement is expected to exceed 2 cm (0.75 in.).

Installation and field inspection procedures for posttensioned slab-on-grade have been prepared by the Post-Tensioning Institute (1996). Posttensioned slab-on-grade consists of concrete with embedded steel tendons that are encased in thick plastic sheaths. The plastic sheath prevents the tendon from coming in contact with the concrete and permits the tendon to slide within the hardened concrete during the tensioning operations. Usually tendons have a dead end (anchoring plate) in the perimeter (edge) beam and a stressing end at the opposite perimeter beam to enable the tendons to be stressed from one end. However, the Post-Tensioning Institute (1996) does recommend that the tendons in excess of 30 m (100 ft) be stressed from both ends. The Post-Tensioning Institute also provides typical anchorage details for the tendons.

The purpose of this section is to describe the settlement behavior of posttensioned slab-on-grade. Two cases (A and B) of posttensioned slab-on-grade will be discussed.

Case A: Posttensioned slab-on-grade. The first case study deals with fill settlement that affected a posttensioned slab-on-grade for a single-family house having an attached garage located in Oceanside, California. The house was constructed in 1987 using typical wood-frame construction with interior wallboard and an exterior stucco facade. A posttensioned slab-on-grade was recommended because the site had alluvium and colluvium that could not be removed because of the presence of a shallow groundwater table.

The posttensioned slab-on-grade was not designed as a "ribbed" (or waffle) type foundation having stiffening beams projecting from the bottom of the slab in both directions. Instead, the posttensioned slab-on-grade was designed so that it consisted of a uniform-thickness slab with an edge beam at the entire perimeter, but no intersecting interior stiffening beams. This type of posttensioned slab-on-grade is commonly referred to as the *California slab* or the *California foundation* (Post-Tensioning Institute, 1996).

The posttensioned slab-on-grade consisted of an 0.5-m- (18-in.-) thick perimeter edge beam and a 13-cm- (5-in.-) thick slab. The edge beam and slab were placed at the same time in order to create a monolithic foundation. The posttensioning tendons were 1.3-cm- (0.5-in.-) diameter cables having seven strands and an ultimate capacity of 1.9 GPa (270 ksi). The tendons were spaced 1.7 m (5.5 ft) on center (both ways) and each tendon was tensioned with a force of approximately 110 kN (25 kips). Based on the tendon spacing, tendon force, and thickness of slab, the compressive stress in the slab due to the posttensioning is about 0.5 MPa (75 psi).

Figure 4.22 presents a manometer survey performed on the posttensioned slab-on-grade. Including measurements taken on the attached garage stemwall, the maximum slab differential Δ is about 7.6 cm (3.0 in.) and the maximum angular distortion δ/L is about 1/120. If the maximum slab differential Δ of 7.6 cm (3.0 in.) versus maximum angular distortion δ/L of 1/120 is plotted in Fig. 4.2, the data point is consistent with the data from other types of foundations. As shown in Fig. 4.22, there is a distinct tilt in the posttensioned

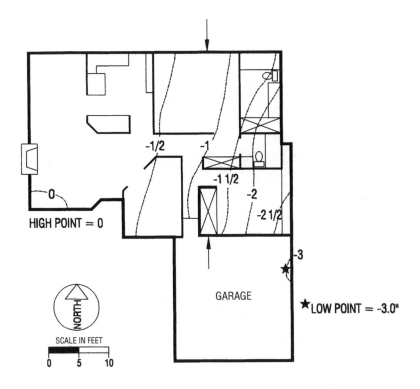

CONTOURS ARE OF RELATIVE ELEVATION IN INCHES

Figure 4.22 Manometer survey (Case A). [*From Day (1998b), reprinted with permission from the American Society of Civil Engineers.*]

slab-on-grade. It was observed that about one-half of the posttensioned slab-on-grade tilted downward in the direction of deepening fill. There was up to about 3 m (10 ft) of poorly compacted fill that settled beneath the east side of the house and caused the foundation displacement shown in Fig. 4.22.

Figure 4.23 is a photograph of the observed posttensioned slab crack. The slab crack was relatively linear and the location is indicated by the two arrows on Fig. 4.22. The slab crack appeared to act as a hinge point, with the most significant wallboard and ceiling cracks located in an area approximately parallel to the slab crack. Considering the magnitude of foundation displacement, the width of the crack was relatively small (1.5 mm, 0.06 in.). This was attributed to the posttensioning compression effect, which prevented the foundation from pulling apart when it settled.

Figure 4.23 View of posttensioned slab crack. [*From Day (1998b), reprinted with permission from the American Society of Civil Engineers.*]

Although the slab crack was relatively small (Fig. 4.23), there was significant cracking to the house including wallboard, ceiling, and exterior stucco cracks. On the basis of the observed interior and exterior cracking to the house, and using Table 4.2 as a guide, the damage could be classified as "moderate," with functional (or serviceability) damage.

In summary, the width of the slab crack was relatively small due to the posttensioning (compression) effect on the slab. However, because of the development of a hinge point, the posttensioned slab deformed to such an extent that there was moderate damage to the interior and exterior wood-framed walls.

Case B: Posttensioned slab-on-grade. Case B is similar to Case A in that the site consists of a single-family house having a posttensioned slab-on-grade (California slab) that was affected by settlement. The site is located in San Clemente, California. The site originally consisted of a deep canyon which required substantial cut-and-fill operations to create a level building pad. The original geotechnical engineer for the site recognized that there would be postconstruction settlement of the fill due to its own weight and variations in moisture content. Given the likelihood of settlement of the deep canyon fill, in 1985 the geotechnical engineer recommended that the structural engineer design a posttensioned slab-on-grade that would be compatible with a maximum angular distortion δ/L on the order of 1/300. For a fill thickness of 18 m (60 ft) the geotechnical engineer estimated a long-term fill settlement ρ_{max} of 5 cm (2 in.).

As in Case A, the house was built using wood-frame construction with interior wallboard and an exterior stucco facade. Also, the construction details for the posttensioned slab (California slab) were similar for both cases. For example, the slab thickness was 13 cm (5 in.), there were perimeter edge beams, and the tendon spacing was 1.5 m (5 ft) on center, both ways.

Figure 4.24 presents the manometer survey performed on the posttensioned slab-on-grade. Including measurements taken on the attached garage stemwall, the maximum slab differential Δ is about 12 cm (4.7 in.) and the maximum angular distortion δ/L is about 1/115. If the maximum slab differential Δ of 12 cm (4.7 in.) versus maximum angular distortion δ/L of 1/115 is plotted in Fig. 4.2, the data point is consistent with the data from other types of foundations. Elevation surveys of the top of curbs indicated that the site had indeed settled and that the deformation shown in Fig. 4.24 represented actual downward movement of the house foundation.

As shown in Fig. 4.24, there is a distinct tilt in the posttensioned slab-on-grade. It was observed that the posttensioned slab-on-grade tilted downward in the direction of deepening fill. In Fig. 4.24, the depths of fill have been indicated at the corners of the house. There is about 7 m (23 ft) of fill at the southeast corner of the house, and the fill uniformly increases to about 19 m (61 ft) at the northwest corner of the house. Results of the subsurface exploration and laboratory testing indicated that the fill settlement was mainly restricted to the deeper zones of fill. The main process of fill settlement was a result of infiltration of water into the fill which caused soil collapse (Lawton et al., 1989, 1991, and 1992).

As indicated in Table 4.2, a maximum differential settlement Δ of 12 cm (4.7 in.) and a maximum angular distortion δ/L of 1/115 for the

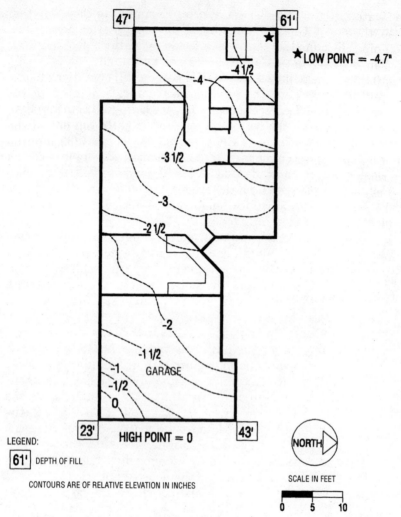

Figure 4.24 Manometer survey (Case B). [*From Day (1998b), reprinted with permission from the American Society of Civil Engineers.*]

foundation should result in "severe" damage. But, based on observed damage, the damage classification was only "slight" and consisted of architectural damage. For example, cracking of the house included minor interior wallboard cracks and a few exterior stucco cracks at the corners of window and door openings. There was a 0.8-mm- (0.03-in.-) wide crack in the garage portion of the posttensioned slab-on-grade. The most significant damage was to exterior utilities and appurtenances. For example, there was a pipe leak under the sidewalk located at the front of the house. When the pipe break was exposed, there was apparently an 8-cm (3-in.) offset of the ends of the broken pipe. In addition, there were separations on the order of 1.3 cm (0.5 in.) between the concrete driveway and the house and up to 0.8-cm (0.3-in.) separation of the rear patio from the house.

As previously mentioned, these tension-type features are common for fill settlement over the sidewalls of a filled-in canyon (Fig. 4.5). Because of the posttensioning (compression) of the foundation at the site, the tension effects were not present and mainly developed in the exterior appurtenances.

The reason that the damage was only slight was due to the posttensioned slab-on-grade which was able to resist the tension effects of the fill settlement. Also, a hinge point did not develop in the slab. Instead, there was a tendency for the posttensioned slab-on-grade to tilt uniformly, similar to rigid body movement, in the direction of deepening fill.

Summary. Cases A and B indicate that the amount of damage to the structure depends on how the posttensioned slab-on-grade deforms. For Case A, the width of the slab crack (Fig. 4.23) was relatively small due to the posttensioning (compression) effect of the slab which prevented the foundation from pulling apart when it settled. However, the posttensioned slab nevertheless deformed to such an extent that there was moderate damage to the interior and exterior walls. The slab crack appeared to act as a hinge point, with the most significant wallboard and ceiling cracks located in an area approximately parallel to the slab crack. The hinge point developed because about half of the posttensioned slab-on-grade was unaffected by soil movement and remained relatively level, while the remaining half of the slab settled. For this case study, the Skempton and MacDonald (1956) criteria that cracking is likely to occur if the angular distortion of the foundation exceeds 1/300 seems reasonable. Also, Table 4.2, which correlates the damage category with crack widths, maximum differential settlement Δ, and maximum angular distortion δ/L, was applicable.

Once again, for Case B, the width of the slab crack was relatively small due to the posttensioning (compression) effect which enabled the slab to resist the tensional forces caused by fill settlement over the sidewall of a preexisting canyon. For Case B, the posttensioned slab-on-grade was able to adjust to the deep settlement. For example, there was a tendency for the posttensioned slab-on-grade to tilt uniformly (no hinge point). Table 4.2 did not accurately correlate the damage category with crack widths and maximum differential settlement Δ for the posttensioned slab-on-grade (California slab) subjected to foundation displacement for Case B.

These case studies indicate the importance of the foundation type (conventional lightly reinforced versus posttensioned slab-on-grade) and mode of foundation deformation (development of a hinge point) in the magnitude of damage to the structure.

5

Expansive Soil

The following notation is introduced in this chapter:

Symbol	Definition
e_m	Edge moisture variation distance
EWL	Equivalent wheel loads
G_f	Gravel equivalent factor
N	Vertical pressure for one-dimensional swell test
PL	Plastic limit
R	R-value
SL	Shrinkage limit
T	Flexible pavement thickness
TI	Traffic index
w	Water content or moisture content
y_m	Maximum anticipated vertical differential movement

5.1 Introduction

Expansive soils are a worldwide problem, causing extensive damage to civil engineering structures. Jones and Holtz estimated in 1973 that the annual cost of damage in the United States due to expansive soil movement was $2.3 billion (Jones and Holtz, 1973). Although most states have expansive soil, Chen (1988) reported that certain areas of the United States, such as Colorado, Texas, Wyoming, and California, are more susceptible to damage from expansive soils than

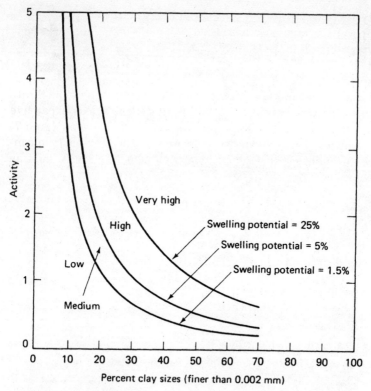

Figure 5.1 Classification chart for swelling potential. (*From Seed et al., 1962, reprinted with permission from the American Society of Civil Engineers.*)

others. These areas have large surface deposits of clay and have climates characterized by alternating periods of rainfall and drought.

5.1.1 Expansive soil factors

There are many factors that govern the expansion behavior of soil. The primary factors are the availability of moisture and the amount and type of the clay particles in the soil. For example, Seed et al. (1962) developed a classification chart based solely on the amount and type (activity) of clay particles (see Fig. 5.1). Other factors affecting the expansion behavior include the type of soil (natural or fill), the condition of the soil in terms of dry density and moisture content, the magnitude of the surcharge pressure, the amount of nonexpansive material (gravel or cobble-size particles), and the amount of aging (Ladd and Lambe, 1961; Kassiff and Baker, 1971; Chen, 1988; Day, 1991b, 1992a). In general, expansion potential increases as the dry density increases and the moisture content decreases. Also, the expansion potential increases as the surcharge pressure decreases.

As shown in Fig. 5.1, the more clay particles of a particular type a soil has, the more swell there will be (all other factors being the same). Likewise, as shown in Fig. 5.1, the type of clay particles significantly affects swell potential. Given the same dry weight, kaolinite clay particles (activity between 0.3 and 0.5) are much less expansive than sodium montmorillonite clay particles (activity between 4 and 7) (Holtz and Kovacs, 1981). Using such factors as the clay-particle content, Holtz and Gibbs (1956) developed a system to classify soils as having either a low, medium, high, or very high expansion potential. Table 5.1 lists typical soil properties versus the expansion potential (Holtz and Gibbs, 1956; *Uniform Building Code,* 1997; Holtz and Kovacs, 1981; and Meehan and Karp, 1994).

5.1.2 Laboratory testing

The forensic engineer can determine the presence of expansive soil by performing subsurface exploration and laboratory testing. One common laboratory test used to determine the expansion potential of the soil is the expansion index test. The test provisions are stated in the *Uniform Building Code* (1997), titled "Uniform Building Code Standard 18-2, Expansion Index Test," and in ASTM (1997k), which has a nearly identical test specification (D 4829). The purpose of this laboratory test is to determine the expansion index, which is then used to classify the soil as having either a very low, low, medium, high, or very high expansion potential as shown in Table 5.1.

TABLE 5.1 Typical Soil Properties versus Expansion Potential

Expansion Potential (1)	Very low (2)	Low (3)	Medium (4)	High (5)	Very high (6)
Expansion index	0–20	21–50	51–90	91–130	130+
Clay content (<2 μm), %	0–10	10–15	15–25	25–35	35–100
Plasticity index	0–10	10–15	15–25	25–35	35+
% swell at 2.8 kPa (60 psf)	0–3	3–5	5–10	10–15	15+
% swell at 6.9 kPa (144 psf)	0–2	2–4	4–7	7–12	12+
% swell at 31 kPa (650 psf)	0	0–1	1–4	4–6	6+

NOTE: % swell for specimens at moisture and density conditions per U.S. Department of Housing and Urban Development (HUD) criteria.

Other laboratory tests, such as hydrometer analyses and Atterberg limits, can be used to classify the soil and estimate its expansiveness. Those soils having a high clay content and plasticity index, such as clays of high plasticity (CH), will usually be classified as having a high to very high expansion potential (see Table 5.1).

The most direct method of determining the amount of swelling is to perform a one-dimensional swell test by utilizing the oedometer apparatus. The undisturbed soil specimen is placed in the oedometer and a vertical pressure (also referred to as *surcharge pressure*) is applied. Then the soil specimen is inundated with distilled water and the one-dimensional vertical swell is calculated as the increase in height of the soil specimen divided by the initial height, often expressed as a percentage. Such a test offers an easy and accurate method of determining the percent swell of the soil. After the soil has completed its swelling, the vertical pressure can be increased to determine the swelling pressure, which is defined as that pressure required to return the soil specimen to its original (initial) height (Chen, 1988).

5.1.3 Surcharge pressure

The bottom three rows of Table 5.1 list typical values of percent swell versus expansion index. Note in Table 5.1 the importance of surcharge pressure on percent swell. At a surcharge pressure of 31 kPa (650 psf) the percent swell is much less than at a surcharge pressure of 2.8 kPa (60 psf). For example, for "highly" expansive soil, the percent swell for a surcharge pressure of 2.8 kPa (60 psf) is typically 10 to 15 percent, while at a surcharge pressure of 31 kPa (650 psf), the percent swell is 4 to 6 percent. The effect of surcharge is important because it is usually the lightly loaded structures, such as concrete flatwork, pavements, slab-on-grade foundations, or concrete canal liners that are significantly impacted by expansive soil.

5.2   Swelling of Desiccated Clay

Desiccated clays are common in areas having near-surface clay deposits and periods of drought. Structures constructed on top of desiccated clay can be severely damaged due to expansive soil heave (Jennings, 1953). For example, Chen (1988) states: "Very dry clays with natural moisture content below 15 percent usually indicate danger. Such clays will easily absorb moisture to as high as 35 percent with resultant damaging expansion to structures." There can also be desiccation and damage to final clay cover systems for landfills and site remediation projects, and for shallow clay landfill liners (Boardman and Daniel, 1996).

Figure 5.2 Ground surface cracks associated with a desiccated clay deposit.

5.2.1   Identification of desiccated clay

The forensic engineer can often visually identify desiccated clay because of the numerous ground surface cracks, such as shown in Fig. 5.2. Desiccated clay deposits also generally have a distinct water content profile, where the water content increases with depth. For example, Fig. 5.3 shows the water content versus depth for two clay deposits located in Irbid, Jordan (Al-Homoud et al., 1997). Soil deposit A has a liquid limit of 35 and a plasticity index of 22, while soil deposit B has a liquid limit of 79 and a plasticity index of 27 (Al-Homoud et al., 1995).

Note in Fig. 5.3 that during the hot and dry summer, the water content of the soil is significantly lower than during the wet winter. During the summer, the lowest water contents are recorded near ground surface, and the water contents are below the shrinkage limit (SL). A near-surface water content below the shrinkage limit is indicative of severe desiccation of the clay.

Below a depth of about 3.2 m (10 ft) for soil deposit A and a depth of about 4.5 m (15 ft) for soil deposit B, the water content is relatively unchanged between the summer and winter monitoring period, and this depth is commonly known as the *depth of seasonal moisture change*. The depth of seasonal moisture change would depend on many factors such as the temperature and humidity, length of the drying season, presence of vegetation that can extract soil moisture, depth of the groundwater table, and the nature of the soil in terms of

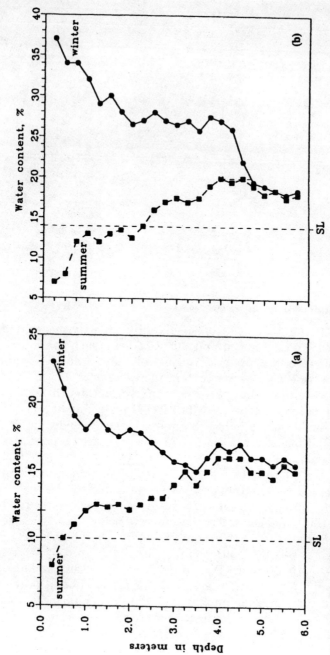

Figure 5.3 Change in water content for: (a) soil A; (b) soil B. (*From Al-Homoud et al., 1997, reprinted with permission from the American Society of Civil Engineers.*)

Figure 5.4 Specimen of desiccated Otay Mesa clay.

clay content, clay mineralogy, and density. As shown in Fig. 5.3, soil
deposit B has a greater variation in water content from the dry summer to wet winter and a greater depth of seasonal moisture change.
This is probably because soil deposit B has a higher clay content than
soil deposit A.

5.2.2 Hydraulic conductivity and rate of swell

Figure 5.4 shows a specimen of desiccated natural clay obtained from
Otay Mesa, California. The near-surface deposit of natural clay has
caused extensive damage to structures, pavements, and flatwork constructed in this area. The clay particles in this soil are almost exclusively montmorillonite (Kennedy and Tan, 1977; Cleveland, 1960).

Figure 5.5 presents the results of a one-dimensional swell test
(lower half of Fig. 5.5) and a falling head permeameter test (upper
half of Fig. 5.5) performed on a specimen of desiccated Otay Mesa clay
(Day, 1997b). At time zero, the desiccated clay specimen was inundated with distilled water. The data in Fig. 5.5 indicates three separate phases of swelling of the clay, as follows:

1. *Primary swell.* The first phase of swelling of the desiccated
clay was primary swell. The primary swell occurs from time equals

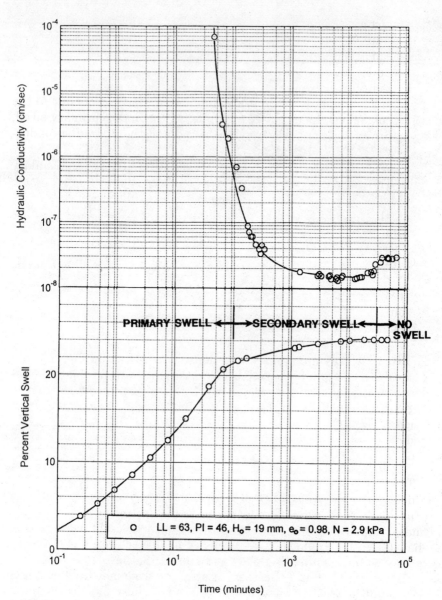

Figure 5.5 Hydraulic conductivity and percent swell versus time.

zero (start of wetting) to about 100 minutes. The end of primary swell (100 minutes) was estimated from Casagrande's log-time method, which can also be applied to the swelling of clays (Day, 1992b).

Figure 5.5 shows that during primary swell, there was a rapid decrease in the hydraulic conductivity (also known as *permeability*) of the clay. The rapid decrease in hydraulic conductivity was due to the closing of soil cracks as the clay swells. At the end of primary swell, the main soil cracks have probably closed and the hydraulic conductivity was about 7×10^{-7} cm/s.

2. *Secondary swell.*   The second phase of swelling was secondary swell. The secondary swell occurs from a time of about 100 minutes to 20,000 minutes after wetting. Figure 5.5 shows that during secondary swell, the hydraulic conductivity continues to decrease as the clay continues to swell and the microcracks close up. The lowest hydraulic conductivity of about 1.5×10^{-8} cm/s occurs at a time of about 5000 minutes, when most of the microcracks have probably sealed up. From a time of 5000 to 20,000 minutes after wetting, there was a slight increase in hydraulic conductivity. This is probably due to a combination of additional secondary swell which increases the void ratio and a reduction in entrapped air.

3. *Steady state.*   The third phase started when the clay stopped swelling. This occurred at about 20,000 minutes after inundation with distilled water. No swell was recorded from a time of 20,000 minutes after wetting to the end of the test (50,000 minutes). As shown in Fig. 5.5, the hydraulic conductivity is constant once the clay has stopped swelling. From a time of 20,000 minutes after wetting to the end of the test (65,000 minutes), the hydraulic conductivity of the clay was constant at about 3×10^{-8} cm/s.

The shape of the swelling versus time curve (Fig. 5.5, lower half) of a desiccated clay is similar to the consolidation of a saturated clay (Fig. 4.14), except that swelling has been plotted as positive values and consolidation as negative values. There appear to be three factors that govern the rate of swelling of desiccated clay:

1. *Development of desiccation cracks.*   The amount and distribution of desiccation cracks, such as shown in Figs. 5.2 and 5.4, are probably the greatest factors in the rate of swelling. Clays will shrink until the shrinkage limit (usually a low water content) is reached. Even as the moisture content decreases below the shrinkage limit, there is probably still the development of additional microcracks as the clay dries. The more cracks in the clay, the greater the pathways for water to penetrate the soil, and the quicker the rate of swelling.

2. *Increased suction at a lower water content.* The second factor that governs the rate of swelling of a desiccated clay is suction pressure. It is well known that, as the water content decreases, the suction pressure of the clay increases (Fredlund and Rahardjo, 1993). At low water contents, the water is drawn into the clay by the suction pressures. The combination of both shrinkage cracks and high suction pressures allows water to be quickly sucked into the clay, resulting in a higher rate of swell.

3. *Slaking.* The third reason is the process of slaking. *Slaking* is defined as the breaking of dried clay when submerged in water, due either to compression of entrapped air by inwardly migrating capillary water or to the progressive swelling and sloughing off of the outer layers (Stokes and Varnes, 1955). Slaking breaks apart the dried clay clods and allows water to quickly penetrate all portions of the desiccated clay. The process of slaking is quicker and more disruptive for clays having the most drying time and lowest initial water content.

In summary, a structure that is constructed during a hot and dry season when the near-surface water content is below the shrinkage limit, such as shown in Fig. 5.3, would be most susceptible to damage. A severely desiccated near-surface clay will have the greatest potential increase in water content and resulting heave of the structure. The hydraulic conductivity (permeability) and rate of swell are important because they determine how fast the water will penetrate the soil.

5.3 Types of Expansive Soil Movement

5.3.1 Lateral movement

Expansive soil movement can affect all types of civil engineering projects. For example, Fig. 5.6 shows the collapse of a retaining wall due to the pressures from an expansive soil backfill. For many retaining and basement walls, especially if the clay backfill is compacted below optimum moisture content, seepage of water into the clay backfill causes horizontal swelling pressures well in excess of at-rest values. Fourie (1989) measured the swell pressure of a compacted clay for zero lateral strain to be 420 kPa (8800 psf). Besides the swelling pressure induced by the expansive soil, there can also be groundwater or perched water pressure on the retaining or basement wall because of the poor drainage of clayey soils. Because of these detrimental effects of clay backfill, a common recommendation is to use only free-drain-

Figure 5.6 Collapse of a retaining wall that has a clay backfill.

ing, granular material (sand or gravel) as retaining or basement wall backfill.

5.3.2 Vertical movement

If a structure having a large area, such as a pavement or foundation, is constructed on top of a desiccated clay, there are usually two main types of expansive soil movement:

1. *Cyclic heave and shrinkage.* The first type of expansive soil movement is cyclic heave and shrinkage which commonly affects the perimeter of the foundation, by uplifting the edge of the structure or shrinking away from it.

Clays are characterized by their moisture sensitivity. They will expand when given access to water and shrink when they are dried out. A soil classified as having a "very high" expansion potential will swell or shrink much more than a soil classified as having a "very low" expansion potential (Table 5.1). For example, the perimeter of a pavement or slab-on-grade foundation will heave during the rainy season and then settle during the drought if the clay dries out. This causes cycles of up and down movement, causing cracking and damage to the structure. Field measurements of this up-and-down cyclic movement have been recorded by Johnson (1980).

The amount of cyclic heave and shrinkage depends on the change in moisture content of the clays below the perimeter of the structure. The moisture change in turn depends on the severity of the drought and rainy seasons, the influence of drainage and irrigation, and the presence of live tree roots, which can extract moisture and cause clays to shrink. The cyclic heave and shrinkage around the perimeter of a structure is generally described as a seasonal or short-term condition.

2. *Progressive swelling beneath the center of the structure (center lift)*. Two ways that moisture can accumulate underneath structures are by thermal osmosis and capillary action. It has been stated that water at a higher temperature than its surroundings will migrate in the soil toward the cooler area to equalize the thermal energy of the two areas (Chen, 1988; Nelson and Miller, 1992). This process has been termed "thermal osmosis" (Sowers, 1979; Day, 1996a). Especially during the summer months, the temperature under the center of a structure tends to be much cooler than at the exterior ground surface.

Because of capillary action, moisture can move upward through soil, where it will evaporate at the ground surface. But when a structure is constructed, it acts as a ground surface barrier, reducing or preventing the evaporation of moisture. It is the effect of thermal osmosis and the evaporation barrier due to the structure that causes moisture to accumulate underneath the center of the structure. A moisture increase will result in swelling of expansive soils. The progressive heave of the center of the structure is generally described as a long-term condition, because the maximum value may not be reached until many years after construction. Figure 5.7 illustrates center lift beneath a house foundation and Fig. 5.8 shows the typical crack pattern in the concrete slab-on-grade due to expansive soil center lift.

5.4   Foundation Design for Expansive Soil

5.4.1   Conventional slab-on-grade foundation

A conventional slab-on-grade consists of interior and exterior bearing wall footings and an interior slab. The concrete for the footings and slab is usually placed at the same time to create a monolithic foundation. For the conventional slab-on-grade foundation, the soils engineer rather than the structural engineer usually provides the construction details. The purpose of the design is to provide deepened perimeter footings that are below the primary zone of cyclic heave and shrinkage and to soak the subgrade soils (prior to construction)

HEAVE
Swell from water
introduction

SETTLEMENT
Shrinkage from
water extraction

broken slab

active zone

roots

EXPANSIVE SOIL MOVEMENT

Figure 5.7   Center lift of a house foundation.

in order to reduce the long-term progressive swelling beneath the center of the slab.

Table 5.2 presents an example of expansive soil foundation specifications for conventional slab-on-grade on expansive soils in southern California (Day, 1994b). Table 5.2 indicates that both the depth of perimeter footings and the depth of presaturation should be increased as the soil becomes more expansive.

Problems can develop because the conventional slab-on-grade is designed, through the use of deepened footings and presoaking, to reduce the effects of expansive soil forces rather than make the foundation strong enough to resist such forces. Two of the most common problems are (1) the perimeter footings are not deep enough to resist the cyclic heave and shrinkage and (2) the presoaking beneath the slab is not done properly (the soil is not completely saturated), leaving the potential for long-term progressive swelling beneath the center of the slab. Figure 5.9 shows a manometer survey on a conventional slab-on-grade that experienced damage due to progressive expansive soil heave caused by moisture migration to the center of the covered area. At this site, the soil was classified as "very highly" expansive (Table 5.1).

5.4.2   Posttensioned slab-on-grade

A second type of foundation for expansive soils is the posttensioned slab-on-grade. There can be many different types of posttensioned

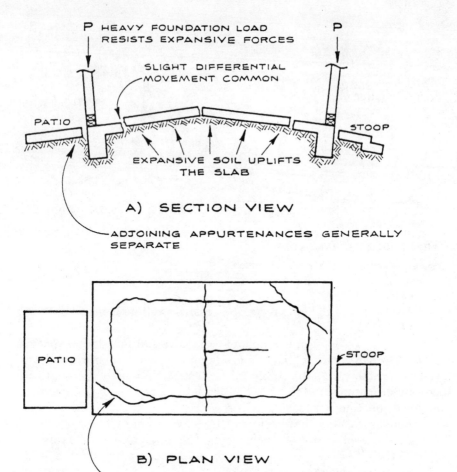

P HEAVY FOUNDATION LOAD
RESISTS EXPANSIVE FORCES

SLIGHT DIFFERENTIAL
MOVEMENT COMMON

PATIO

STOOP

EXPANSIVE SOIL UPLIFTS
THE SLAB

A) SECTION VIEW

ADJOINING APPURTENANCES GENERALLY
SEPARATE

PATIO

STOOP

B) PLAN VIEW

CRACKS TEND TO PARALLEL WALLS
AND RUN DIAGONALLY ACROSS CORNERS.
CRACKS ALSO FREQUENTLY OCCUR ACROSS
THE CENTRAL PORTION OF THE SLAB SECTION.

Figure 5.8 Typical crack pattern due to center lift.

designs. In Texas and Louisiana, the early posttensioned foundations consisted of a uniform-thickness slab with stiffening beams in both directions, which became known as the *ribbed foundation*. As previously mentioned, in California a commonly used type of posttensioned slab consists of a uniform thickness slab with an edge beam at the entire perimeter, but no or minimal interior stiffening beams or ribs.

TABLE 5.2   **Expansive Soil Foundation Recommendations**

Expansive classification (1)	Depth of footing below adjacent grade (2)	Footing reinforcement (3)	Slab thickness and reinforcement conditions (4)	Presaturation below slabs (5)	Rock base below slabs (6)
None to low	0.5 m (18 in.) exterior 0.3 m (12 in.) interior	Exterior: 4 #4 bars, 2 top and 2 bottom Interior: 2 #4 bars, 1 top and 1 bottom	0.10 m (4 in.) nominal with #3 bars at 0.4 m (16 in.) on center, each way	To 0.3 m (12 in.)	Optional
Medium	0.6 m (24 in.) exterior 0.5 m (18 in.) interior	Exterior: 4 #5 bars, 2 top and 2 bottom Interior: 4 #4 bars, 2 top and 2 bottom	0.10 m (4 in.) net with #3 bars at 0.3 m (12 in.) on center, each way	To 0.5 m (18 in.)	0.10 m (4 in.)
High	0.8 m (30 in.) exterior 0.5 m (18 in.) interior	Exterior: 4 #5 bars, 2 top and 2 bottom Interior: 4 #5 bars, 2 top and 2 bottom	0.13 m (5 in.) net with #4 bars at 0.4 m (16 in.) on center, each way	To 0.6 m (24 in.)	0.15 m (6 in.)
Very high	0.9 m (36 in.) exterior 0.8 m (30 in.) interior	Exterior: 6 #5 bars, 3 top and 3 bottom Interior: 4 #5 bars, 2 top and 2 bottom	0.15 m (6 in.) nominal with #4 bars at 0.3 m (12 in.) on center, each way	To 0.8 m (30 in.)	0.20 m (8 in.)

This type of posttensioned slab has been termed the *California slab* (Post-Tensioning Institute, 1996).

In "Design and Construction of Post-Tensioned Slabs-on-Ground," prepared by the Post-Tensioning Institute (1996), the design moments, shears, and differential deflections under the action of soil loading resulting from changes in moisture contents of expansive soils are predicted using equations developed from empirical data and a computer study of a plate on an elastic foundation. *The Uniform Building Code* (1997) presents nearly identical equations for the design of posttensioned slabs-on-grade. The idea for the design of a posttensioned slab-on-grade in accordance with the Post-Tensioning Institute is to construct a slab foundation that is strong enough and rigid enough to resist the expansive soil forces. To get the required rigidity to reduce foundation deflections, the stiffening beams

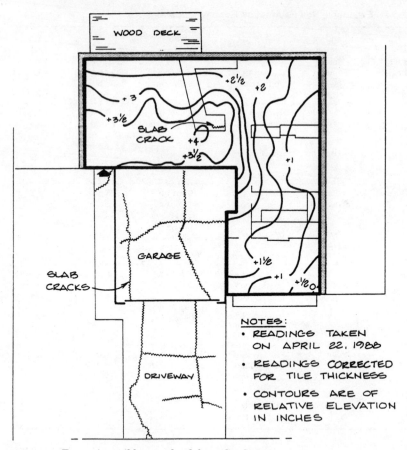

Figure 5.9 Expansive soil heave of a slab-on-Grade.

(perimeter and interior footings) can be deepened. Although the differential movement to be expected for a given expansive soil is supplied by a soils engineer, the actual design of the foundation is usually by the structural engineer.

The posttensioned slab-on-grade should be designed for two conditions: (1) center lift (also called *center heave* or *doming*) and (2) edge lift (also called *edge heave* or *dishing*). Center lift is a result of long-term progressive swelling beneath the center of the slab, or drying and shrinking of the soil around the perimeter of the slab (causing the perimeter to settle), or a combination of both. Edge lift is the cyclic heave beneath the perimeter of the foundation. In order to complete the design, the soils engineer usually provides the maximum anticipated vertical differential soil movement (y_m) and the horizontal distance of moisture variation from the slab perimeter (e_m) for both the center lift and edge lift conditions.

There are many different problems that can result in damage to posttensioned slabs-on-grade, and three of the most common are as follows:

1. The design of the foundation is based on static values of y_m, but the actual movement is cyclic. This can cause utility lines that enter the slab to fracture or cracks to develop in interior wallboard due to cyclic movement.

2. The values of e_m are difficult to determine because they are dependent on soil and structural interaction. The structural parameters that govern e_m include the magnitude and distribution of dead loads, the rigidity of the foundation, and the depth of the perimeter footings. The soil parameters include the amount and specific limits of the heave and shrinkage.

3. The soils engineer might test the expansive clays during the rainy season, but the foundation might be built during a drought. In this case, the values of y_m for center lift could be considerably underestimated.

5.4.3 Pier and grade beam support

A third common foundation type for expansive soil is pier and grade beam support, as shown in Fig. 5.10 (from Chen, 1988). The basic principle is to construct the piers such that they are below the depth of seasonal moisture changes. The piers can be belled at the bottom to increase their uplift resistance. As an alternative, it has been stated that the depth of piles or piers should be 1.5 times the depth where the swelling pressure is equal to the overburden pressure (David and Komornik, 1980). Grade beams and structural floor systems that are free of the ground are supported by the piers.

Chen (1988) provides several examples of expansive soil–related damage to pier and grade beam foundations. Common problems include insufficient pier length, excessive pier diameter, absent or inadequate pier reinforcement, excess concrete on the top of the piers, and an absence or inadequate air void space below the grade beams (Woodward et al., 1972; Jubenville and Hepworth, 1981; Chen, 1988).

5.4.4 Other treatment alternatives

Besides the construction of special foundations to resist expansive soils, there are other treatment alternatives as indicated in Table 5.3

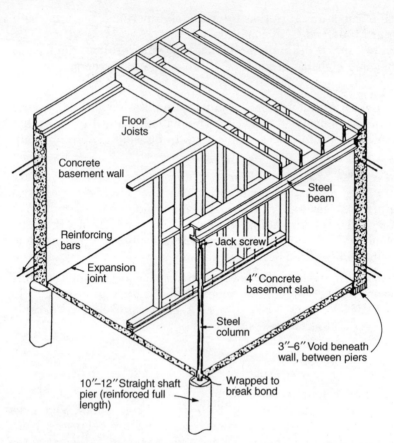

Figure 5.10 Typical detail of the grade beam and pier system. (*From Chen, 1988, reproduced with permission from Elsevier Science—NL.*)

(from Nelson and Miller, 1992). Each of these methods have advantages and disadvantages as indicated in Table 5.3.

5.5   Pavements

5.5.1   Flexible pavements

A typical flexible pavement consists of an aggregate base that is compacted over the subgrade. Asphalt concrete (AC) is used as the wearing surface. Common damage to flexible pavements due to heave of a compacted clay subgrade is cracks in the asphalt concrete that are parallel to the curbs and gutters. This is usually caused by water seeping through the joints in the concrete gutters. The water then

TABLE 5.3   **Expansive Soil Treatment Alternatives**

Method (1)	Salient points (2)
Removal and replacement	Nonexpansive, impermeable fill must be available and economical. Nonexpansive soils can be compacted at higher densities than expansive clay, producing high bearing capacities. If granular fill is used, special precautions must be taken to control drainage away from the fill, so water does not collect in the pervious material. Replacement can provide safe slab-on-grade construction. Expansive material may be subexcavated to a reasonable depth, then protected by a vertical and/or a horizontal membrane. Sprayed asphalt membranes are effectively used in highway construction.
Remolding and compaction	Beneficial for soils having low potential for expansion, high dry density, low natural water content, and in a fractured condition. Soils having a high swell potential may be treated with hydrated lime, thoroughly broken up, and compacted—if they are lime reactive. If lime is not used, the bearing capacity of the remolded soil is usually lower since the soil is generally compacted wet of optimum at a moderate density. Quality control is essential. If the active zone is deep, drainage control is especially important. The specific moisture-density conditions should be maintained until construction begins and checked prior to construction.
Surcharge loading	If swell pressures are low and some deformation can be tolerated, a surcharge load may be effective. A program of soil testing is necessary to determine the depth of the active zone and the maximum swell pressures to be counteracted. Drainage control is important in using a surcharge. Moisture migration can be both vertical and horizontal.
Prewetting	Time periods as long as a year or more may be necessary to increase moisture contents in the active zone. Vertical sand drains drilled in a grid pattern can decrease the wetting time. Highly fissured, desiccated soils respond more favorably to prewetting. Moisture contents should be increased to at least 2–3% above the plastic limit. Surfactants may increase the percolation rate. The time needed to produce the expected swelling may be significantly longer than the time to increase moisture contents. It is almost impossible to adequately prewet dense unfissured clays. Excess water left in the upper soil can cause swelling in deeper layers at a later date. Economics of prewetting can compare favorable to other methods, but funds must be available at an early date in the project. Lime treatment of the surface soil following prewetting can provide a working table for equipment and increase soil strength. Without lime treatment soil strength can be significantly reduced, and the wet surface may make equipment operation difficult. The surface should be protected against evaporation and surface slaking. Quality control improves performance.

TABLE 5.3 (*Continued*) Expansive Soil Treatment Alternatives

Method (1)	Salient points (2)
Lime treatment	Sustained temperatures over 21°C (70°F) for a minimum of 10 to 14 days are necessary for the soil to gain strength. Higher temperatures over a longer time produce higher strength gains. Organics, sulfates, and some iron compounds retard the pozzolanic reaction of lime. Gypsum and ammonium fertilizers can increase the soil's lime requirements. Calcareous and alkaline soil have good reactivity. Poorly drained soils have higher reactivities than well-drained soils. Usually 2–10% lime stabilizes reactive soil. Soil should be tested for lime reactivity and percentage of lime needed. The mixing depth is usually limited to 12–18 in., but large tractors with ripper blades have successfully allowed in-place mixing of 2 ft of soil. Lime can be applied dry or in a slurry, but excess water must be present. Some delay between application and final mixing improves workability and compaction. Quality control is especially important during pulverization, mixing, and compaction. Lime-treated soils should be protected from surface and groundwater. The lime can be leached out and the soil can lose strength if saturated. Dispersion of the lime from drill holes is generally ineffective unless the soil has an extensive network of fissures. Stress relief from drill holes may be a factor in reducing heave. Smaller-diameter drill holes provide less surface area to contact the slurry. Penetration of pressure-injected lime is limited by the slow diffusion rate of the lime, the amount of fracturing in the soil, and the small pore size of clay. Pressure injection of lime may be useful to treat layers deeper than possible with the mixed-in-placed technique.
Cement treatment	Portland cement (4–6%) reduces the potential for volume change. Results are similar to lime, but shrinkage may be less with cement. Method of application is similar to mix-in-place lime treatment, but there is significantly less time delay between application and final placement. Portland cement may not be as effective as lime in treating highly plastic clays. Portland cement may be more effective in treating soils that are not lime-reactive. Higher strength gains may result using cement. Cement-stabilized material may be prone to cracking and should be evaluated prior to use.
Salt treatment	There is no evidence that use of salts other than NaCl or $CaCl_2$ is economically justified. Salts may be leached easily. Lack of permanence of treatment may make salt treatment uneconomical. The relative humidity must be at least 30% before $CaCl_2$ can be used. Calcium and sodium chloride can reduce frost heave by lowering the freezing point of water. $CaCl_2$ may be useful to stabilize soils having a high sulfur content.
Fly ash	Fly ash can increase the pozzolanic reaction of silty soils. The gradation of granular soils can be improved.

TABLE 5.3 (*Continued*) **Expansive Soil Treatment Alternatives**

Method (1)	Salient points (2)
Organic compounds	Spraying and injection are not very effective because of the low rate of diffusion in expansive soil. Many compounds are not water-soluble and react quickly and irreversibly. Organic compounds do not appear to be more effective than lime. None is as economical and effective as lime.
Horizontal barriers	Barrier should extend far enough from the roadway or foundation to prevent horizontal moisture movement into the foundation soils. Extreme care should be taken to securely attach barrier to foundation, seal the joints, and slope the barrier down and away from the structure. Barrier material must be durable and nondegradable. Seams and joints attaching the membrane to a structure should be carefully secured and made waterproof. Shrubbery and large plants should be planted well away from the barrier. Adequate slope should be provided to direct surface drainage away from the edges of the membranes.
Asphalt	When used in highway construction, a continuous membrane should be placed over subgrade and ditches. Remedial repair may be less complex than concrete pavement. Strength of pavement is improved over untreated granular base. Can be effective when used in slab-on-grade construction.
Rigid barrier	Concrete sidewalks should be reinforced. A flexible joint should connect sidewalk and foundation. Barriers should be regularly inspected to check for cracks and leaks.
Vertical barrier	Placement should extend as deep as possible, but equipment limitations often restrict the depth. A minimum of half of the active zone should be used. Backfill material in the trench should be impervious. Types of barriers that have provided control of moisture content are capillary barrier (coarse limestone), lean concrete, asphalt and ground-up tires, polyethylene, and semihardening slurries. A trenching machine is more effective than a backhoe for digging the trench.
Membrane-encapsulated soil layers	Joints must be carefully sealed. Barrier material must be durable to withstand placement. Placement of the first layer of soil over the bottom barrier must be controlled to prevent barrier damage.

SOURCE: Nelson and Miller, 1992.

percolates downward through the base and is absorbed by the compacted clay subgrade, resulting in expansion of the clay. Because the concrete curbs and gutters are stronger and thicker than the asphalt concrete pavement, the asphalt pavement will have more heave relative to the curbs resulting in cracks parallel to the curbs and gutters. Once cracked, the asphalt pavement allows for further infiltration of water, accelerating the heave process. Besides this crack pattern, there can be a variety of heave and crack patterns in pavements on expansive clay subgrade (Van der Merwe and Ahronovitz, 1973; Day, 1995a).

Another effect of moisture infiltration into the compacted clay subgrade is a reduction in the undrained shear strength of the soil. Moisture causes a softening of the compacted clay. The process involves an increase in moisture content and a reduction in negative pore water pressures, which results in a decrease in the undrained shear strength. This makes the subgrade weaker and more susceptible to pavement deterioration, such as alligator cracking and rutting.

The design of flexible pavement in California is usually based on the following equation (California Division of Highways, 1973):

$$T = \frac{0.0032\,TI(100-R)}{G_f} \tag{5.1}$$

where T = flexible pavement thickness (ft)
  TI = traffic index
  R = R-value
  G_f = gravel equivalent factor

This flexible pavement design equation was developed from empirical relationships and the performance of pavements in actual service.

The effect of traffic on the roadway over the design life of the pavement is expressed by the traffic index (TI). The higher the equivalent wheel loads (EWL) during the life of the pavement, the higher the traffic index. Minor residential streets and culs-de-sac have a typical TI of 4, while a major city street or county highway can have a TI of 7 to 9.

The R-value is the resistance of the subgrade or base and refers to the ability of the soil to resist lateral deformation when acted upon by a vertical load. In essence, the R-value is a relative measure of the shear strength of the soil. The R-value is determined by using standardized test procedures and equipment, such as specified by ASTM D 2844-94, "Standard Test Method for Resistance R-value and Expansion Pressure of Compacted Soils" (ASTM, 1997i). The R-value is determined on specimens compacted to anticipated field conditions. Prior to testing, the specimens are soaked in water to achieve saturation. By soaking the soil prior to testing, the R-value is supposed to represent the worst possible state that the soil might attain in the field. A crushed rock base can have an R-value of 75 to 85, while a clay subgrade will usually have an R-value of less than 10.

The gravel equivalent G_f represents the strength factor of the materials in the pavement structural section. The gravel equivalent is an empirical factor developed through research and field experience; it relates the relative strength of a unit thickness of a particular material in terms of an equivalent thickness of gravel. An aggregate base will have $G_f = 1.0$, while a class A cement-treated aggregate base has $G_f = 1.7$.

Equation (5.1) does not have a factor to account for the expansiveness of the subgrade. For example, it is not unusual for a "medium" expansive clay subgrade and a "very highly" expansive clay subgrade to both have an R-value of less than 10. As a result, for both the medium and very highly expansive clay subgrade, the thickness of asphalt concrete and base material are similar given the same traffic index. But because the percent swell is much greater for the very highly expansive soil (see Table 5.1), this pavement will have more uplift and cracking as moisture infiltrates into the clay subgrade. Methods to mitigate the effects of clay expansion include compacting the clay subgrade wet of optimum or adding lime or cement to the clay subgrade (see Table 5.3). When compacting the clay subgrade wet of optimum, a geofabric can be used to prevent the aggregate base from being pushed down into and contaminated by the soft clay subgrade.

When investigating damage to asphalt pavements due to expansive soil movement, the forensic engineer should compare the actual thickness of pavement section to the design values and investigate what (if any) measures were taken during construction to mitigate the expansion potential of the clay subgrade. Another factor that should always be considered by the forensic engineer is the traffic index. In many cases, the traffic index will be underestimated and cracking caused by expansive soil will be aggravated by heavy and unanticipated wheel loads.

5.5.2 Concrete pavements

The same expansive soil mechanisms that affect flexible pavement can also impact concrete pavements. The damage caused by expansive soil may be more severe for concrete pavements than for asphalt pavements, because the concrete is usually more brittle. For example, Fig. 5.11 shows two photographs of concrete pavement distress due to expansive soil. The site shown in Fig. 5.11 is located in Otay Mesa, California. As previously mentioned, this area has extensive expansive soil related damage because the clay has a "very high" expansion potential and the principal clay mineral in the soil is montmorillonite.

Note in Fig. 5.11 that nonuniform movement of the clay subgrade frequently causes differential displacement across the concrete cracks. The opening of cracks allows for infiltration of water, which can lead to more expansion of the clay subgrade and a further widening of the cracks.

There is also the development of a series of concrete cracks, which tend to be parallel to the concrete curbs, as shown in Fig. 5.11. This is frequently caused by cyclic heave and shrinkage of the pavement adjacent the curbs, or by progressive swelling beneath the interior portion of the pavement.

5.6   Flatwork

5.6.1   Upward movement

Flatwork can be defined as appurtenant structures that surround a building or house such as concrete walkways, patios, driveways, and pool decks. It is the lightly loaded structures, such as pavements or lightly loaded foundations, that are commonly damaged by expansive soil. Because flatwork usually supports only its own weight, it can be especially susceptible to expansive soil–related damage such as shown in Fig. 5.12. The arrows in Fig. 5.12 indicate the amount of uplift of the lightly loaded exterior sidewalk relative to the heavily loaded exterior wall of the tilt-up building. The sidewalk heaved to such an extent that the door could not be opened and had to be rehung so that it opened into the building.

Another example is Fig. 5.13, which shows cracking to a concrete driveway due to expansive soil uplift. The expansive soil uplift tends to produce a distinct crack pattern, which has been termed a *spider* or *x-type* crack pattern.

Besides the concrete itself, utilities can also be damaged due to the upward movement of the flatwork. For example, Fig. 5.14 shows a photograph of concrete flatwork that was uplifted due to expansive soil heave. The upward movement of the flatwork bent the utility line and chipped off the stucco as shown in Fig. 5.14.

Figure 5.11   Two views of concrete pavement damage due to expansive soil.

Figure 5.12 Sidewalk uplift due to expansive soil.

Figure 5.13 Driveway cracking due to expansive soil.

Figure 5.14   Concrete flatwork uplifted by expansive soil.

5.6.2   Walking of flatwork

Besides differential movement of the flatwork, there can also be progressive movement of flatwork away from the structure. This lateral movement is known as "walking." Figure 5.15 shows one example of the walking of a concrete sidewalk that has both differential movement due to the expansive soil and a large gap at a sidewalk joint. Another example of walking of flatwork is shown in Fig. 5.16 where a backyard concrete patio slab was built atop highly expansive soils. The patio slab originally abutted the wall of the house, but is now separated from the house by about 4 cm (1.5 in.). At one time, the gap was filled in with concrete, but as Fig. 5.16 shows, the patio slab has continued to walk away from the house. In many cases, there will be appurtenant structures (such as patio shade cover) that are attached to the house and also derive support from the flatwork. As the flatwork walks away from the house, these appurtenant structures are pulled laterally and frequently damaged.

The results of a field experiment indicated that most walking occurs during the wet period (Day, 1992c). The expansion of the clay causes the flatwork to move up and away from the structure. During the dry period, the flatwork does not return to its original position. Then during the next wet period, the expansion of the clay again causes an upward and outward movement of the flatwork. The cycles

Figure 5.15 Walking of concrete sidewalk on expansive soil.

Figure 5.16 Walking of flatwork on expansive soil.

of wetting and drying cause a progressive movement of flatwork away from the building. An important factor in the amount of "walking" is the moisture condition of the clay prior to construction of the flatwork. If the clay is dry, then more initial upward and outward movement can occur during the first wet cycle.

5.7 Case Study

One frequent cause of expansive soil damage is heave of the foundation (center lift) as shown in Fig. 5.7. Foundation movement can also be caused by the shrinkage of clay. In this case study, tree roots and rootlets extracted moisture from the ground, which caused the near surface clay to shrink and the foundation to settle.

There are cases (Cheney and Burford, 1975) where the opposite can also occur, where large trees have been removed and the clay has expanded as the soil moisture increases to its natural state. In the United States, Holtz (1984) states that large, broadleaf, deciduous trees located near the structure cause the greatest changes in moisture and the greatest resulting damage in both arid and humid areas. However, the most dramatic effects are during periods of drought, such as the severe drought in Britain from 1975 to 1976, when the amount of water used by the trees during transpiration greatly exceeded the amount of rainfall within the area containing tree roots. Biddle (1979, 1983) investigated 36 different trees, covering a range of tree species and clay types and concluded that poplars have much greater effects than other species and that the amount of soil movement will depend on the clay shrinkage characteristics. Ravina (1984) states that it is the nonuniform moisture changes and soil heterogeneities that cause the uneven soil movements that damage shallow foundations, structures, and pavements.

Figure 5.17 presents a site plan sketch of two single-story structures at a complex called Barawid Manor, located in San Diego, California. The foundation of the structures consists of shallow perimeter concrete footings that are about 0.3 m (1 ft) deep and isolated interior footings that support a raised wood floor. The structures have typical wood-frame construction and are faced with stucco. As shown in Fig. 5.17, there is a large pepper tree, estimated to be about 9 m (30 ft) tall, located adjacent the structures. Figure 5.18 (view to the east) shows a photograph of the pepper tree. A manometer survey was performed on the wood floor, and the elevation difference was found to be 17 cm (6.7 in.) for building 2 and 8.1 cm (3.2 in.) for building 3. Note in Fig. 5.17 that the low area of the floor is near the pepper tree, while the highest area is at the opposite side of the structure. Associated with the floor differential were interior wallboard cracks, stucco cracks, and founda-

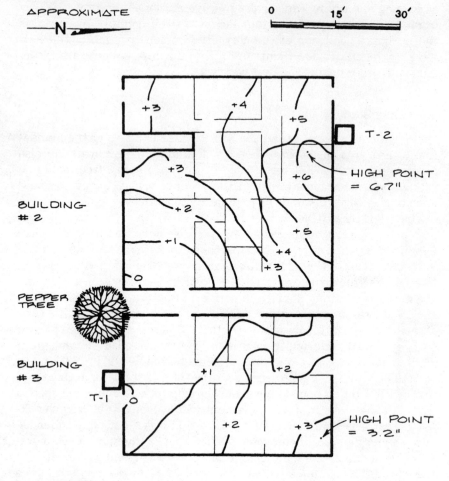

APPROXIMATE

—N—

0   15'   30'

BUILDING #2

T-2

HIGH POINT = 6.7"

PEPPER TREE

BUILDING #3

T-1

HIGH POINT = 3.2"

NOTES:

- READINGS TAKEN ON APRIL 18, 1990
- READINGS CORRECTED FOR TILE THICKNESS
- CONTOURS ARE OF RELATIVE ELEVATION IN INCHES
- MANOMETER PERFORMED ON RAISED WOOD FLOOR

Figure 5.17 Site plan and manometer survey.

Figure 5.18 View of the pepper tree.

tion cracks with the most visible damage on the south side of the build-ings. From Table 4.2, the damage was classified as moderate to severe.

The subsurface exploration consisted of the excavation of two test pits (T-1 and T-2) at the locations shown in Fig. 5.17. The test pits revealed an upper zone of silty sand about 0.2 to 0.3 m (0.7 to 0.9 ft) thick, which was underlaid by clay. The clay was classified as a weathered residual soil. At test pit 1, numerous rootlets and roots up to 5 cm (2 in.) in diameter were observed in the clay.

Particle size analyses indicate that the clay has 19 percent sand, 30 percent silt, and 51 percent clay size particles smaller than 0.002 mm. The soil was tested as having a liquid limit (LL) of 60 and a plas-ticity index (PI) of 42, which would classify the material as a clay of high plasticity (CH). Expansion index tests performed in accordance

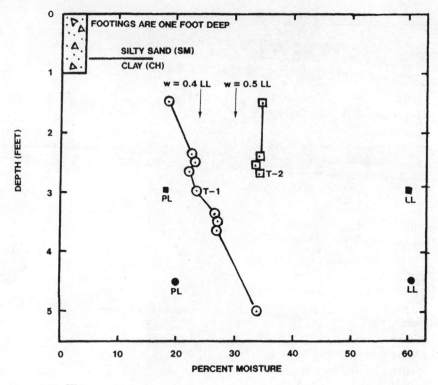

Figure 5.19 Water content versus depth.

with the *Uniform Building Code* (1997) Specification 18-2 indicated an expansion index from 134 to 229, classifying the soil as very highly expansive (Table 5.1).

Figure 5.19 presents the results of moisture content tests performed on undisturbed samples recovered from the test pits. At test pit 1, there was a general increase in moisture content with depth, from 18.8 percent at 0.5 m (1.5 ft) to 33.8 percent at 1.5 m (5 ft). At test pit 2, the moisture contents were about 34 percent. It was concluded that the drier soil at test pit 1 was due to extraction of soil moisture by the pepper tree.

Driscoll (1983) has suggested relationships between moisture content w of the soil and the liquid limit (LL) which indicate the level of desiccation:

$$w = 0.5\text{LL} \tag{5.2}$$

would indicate the onset of desiccation, and

$$w = 0.4\text{LL} \tag{5.3}$$

TABLE 5.4   One-Dimensional Swell Tests

Location (1)	Field dry density, Mg/m^3 (pcf) (2)	Field moisture content, % (3)	Final moisture content, % (4)	Percent swell (5)
T-1 at 0.76 m (2.5 ft)	1.62 (101)	22.5	30.1	4.5
T-1 at 1.07 m (3.5 ft)	1.52 (95)	26.2	33.8	5.2
T-2 at 0.76 m (2.5 ft)	1.38 (86)	34.2	37.8	0.3

NOTE: Swell test performed on undisturbed samples at field moisture and density conditions. For all swell tests, normal pressure is 7 kPa (144 psf).

would indicate the point at which desiccation becomes significant. Given the liquid limit of 60, the water contents for Eqs. (5.2) and (5.3) are plotted in Fig. 5.19. From these relationships, it was concluded that significant desiccation had occurred in the upper 0.6 m (2 ft) of the clay deposit.

Table 5.4 presents the results of one-dimensional vertical swell tests performed on undisturbed specimens. At test pit 1, samples swelled about 5 percent, but at test pit 2, the percent swell was only 0.3 percent. This laboratory testing also indicates the desiccated nature of the clay near the pepper tree.

Tucker and Poor (1978) indicated that trees located at distances closer than their heights to structures caused significantly larger movements due to clay shrinkage than those trees located at greater distances. Hammer and Thompson (1966) stated that trees should not be planted a minimum of one-half their anticipated mature height from a shallow foundation, and slow-growing, shallow-rooted varieties of trees were preferred. However, Cutler and Richardson (1989) indicated that total crown volume (hence leaf area) is generally more important than absolute height in relation to water demand. Cutler and Richardson (1989) used case histories to determine the frequency of damage as a function of tree-trunk distance from the structure for different species. Biddle (1983) recommended that, for very high shrinkage clay, the perimeter footings should be at least 1.5 m (5 ft) deep, and this would be sufficient to accommodate most tree-planting designs. Planting the pepper tree well away from the structures or installing deepened perimeter footings would probably have prevented the damage at this site.

In summary, when dealing with expansive clays, it is important for the forensic engineer to consider the possibility of damage due to clay shrinkage caused by the extraction of moisture by tree roots. For this

Figure 5.20 Expansive soil damage of a raised wood floor foundation.

case study, a large pepper tree caused significant desiccation of the upper 0.6 m (2 ft) of a very highly expansive clay deposit. The shrinkage of the clay deposit caused displacement of the raised wood floor foundation and building damage.

This case study shows that raised wood floor foundations are especially vulnerable to damage due to clay shrinkage. Raised wood floor foundations can also be damaged by clay expansion. For example, Fig. 5.20 is a view from the crawl space of a raised wood floor foundation. At this site, expansive soil has uplifted the lightly loaded concrete pad footing, causing distortion of the wood post supporting the floor beam. Note the wet condition in the crawl space.

6

Lateral Movement

The following notation is introduced in this chapter:

Symbol	Definition
c	Cohesion based on a total stress analysis
c'	Cohesion based on an effective stress analysis
D	Depth of the failure (slip) surface
F	Factor of safety for slope stability
L	Length of the failure slip surface
S	Degree of saturation
ϕ	Friction angle based on a total stress analysis
ϕ'	Friction angle based on an effective stress analysis
γ_b	Buoyant density
γ_t	Wet density
β	Angular distortion as defined by Boscardin and Cording (1989)
α	Slope inclination
ε_h	Horizontal strain of the foundation

6.1 Typical Causes of Lateral Movement

The most common cause of lateral movement of buildings is slope movement. Slope movement can be divided into six basic types, as described below:

1. *Rock falls or topples.* This is usually an extremely rapid movement that includes the free fall of rocks, movement of rocks by leaps and bounds down the slope face, and/or the rolling of rocks or fragments of rocks down the slope face (Varnes, 1978). A rock topple is similar to a rock fall, except that there is a turning

moment about the center of gravity of the rock which results in an initial rotational type movement and detachment from the slope face. Rock falls are discussed in Sec. 6.3.

2. *Surficial slope failure.*   A surficial slope failure involves shear displacement along a distinct failure (or slip) surface. As the name implies, surficial failures develop on the outer face of the slope and are generally shallow [up to 1.2 m (4 ft) deep]. In many cases, the failure surface is parallel to the slope face. Surficial slope failures are discussed in Sec. 6.4.

3. *Gross slope failure.*   As contrasted with a shallow (surficial) failure, a gross slope failure involves shear displacement of the entire slope. Terms such as *fill slope failures* and *earth* or *rock slumps* have been used to identify similar processes. Gross slope failures are discussed in Sec. 6.5.

4. *Landslides.*   The gross failure of a slope could be referred to as a *landslide.* However, landslides in some cases may be so large that they involve several different slopes. Landslides are discussed in Sec. 6.6.

5. *Debris flow.*   Debris flow is commonly defined as soil with entrained water and air that moves readily as a fluid on low slopes. Debris flow can include a wide variety of soil-particle sizes (including boulders) as well as logs, branches, tires, and automobiles. Other terms, such as *mud flow, debris slide, mud slide,* and *earth flow,* have been used to identify similar processes. While categorizing flows based on rate of movement or the percentage of clay particles may be important, the mechanisms of all these flows are essentially the same (Johnson and Rodine, 1984). Debris flow is discussed in Sec. 6.7.

6. *Creep.*   Creep is generally defined as an imperceptibly slow and more or less continuous downward and outward movement of slope-forming soil or rock (Stokes and Varnes, 1955). Creep can affect both the near-surface (surficial) soil or deep-seated (gross) materials. The process of creep is frequently described as viscous shear that produces permanent deformations, but not failure as in landslide movement. Creep is discussed in Sec. 6.8.

   Table 6.1 presents a checklist for the study of slope failures and landslides (adapted from Sowers and Royster, 1978). This table provides a comprehensive list of the factors that may need to be considered by the forensic engineer when investigating slope failures.

   Besides slope movement, there can be many other mechanisms that cause lateral movement of a facility. Examples include the collapse of

TABLE 6.1   Checklist for the Study of Slope Failures and Landslides

Main topic (1)	Relevant items (2)
Topography	Contour map, consider land form and anomalous patterns (jumbled, scarps, bulges).
	Surface drainage, evaluate conditions such as continuous or intermittent drainage.
	Profiles of slope, to be evaluated along with geology and the contour map.
	Topographic changes, such as the rate of change by time and correlate with groundwater, weather, and vibrations.
Geology	Formations at site, consider the sequence of formations, colluvium (bedrock contact and residual soil), formations with bad experience, and rock minerals susceptible to alteration.
	Structure: Evaluate three-dimensional geometry, stratification, folding, strike and dip of bedding or foliation (changes in strike and dip and relation to slope and slide), and strike and dip of joints with relation to slope. Also investigate faults, breccia, and shear zones with relation to slope and slide.
	Weathering: Consider the character (chemical, mechanical, and solution) and depth (uniform or variable).
Groundwater	Piezometric levels within slope, such as normal level, perched levels, or artesian pressures with relation to formations and structure.
	Variations in piezometric levels due to weather, vibration, and history of slope changes. Other factors include response to rainfall, seasonal fluctuations, year-to-year changes, and effect of snowmelt.
	Ground surface indication of subsurface water, such as springs, seeps, damp areas, and vegetation differences.
	Effect of human activity on groundwater, such as groundwater utilization, groundwater flow restriction, impoundment, additions to groundwater, changes in ground cover, infiltration opportunity, and surface water changes.
	Groundwater chemistry, such as dissolved salts and gases and changes in radioactive gases.
Weather	Precipitation from rain or snow. Also consider hourly, daily, monthly, or annual rates.
	Temperature, such as hourly and daily means or extremes, cumulative degree-day deficit (freezing index), and sudden thaws.
	Barometric changes.
Vibration	Seismicity, such as seismic events, microseismic intensity, and microseismic changes.
	Human-induced from blasting, heavy machinery, or transportation (trucks, trains, etc.).
History of slope changes	Natural processes, such as long-term geologic changes, erosion, evidence of past movement, submergence, or emergence.
	Human activities, including cutting, filling, clearing, excavation, cultivation, paving, flooding, and sudden drawdown of reservoirs. Also consider changes caused by human activities, such as changes in surface water, groundwater, and vegetation cover.
	Rate of movement from visual accounts, evidence in vegetation, evidence in topography, or photographs (oblique, aerial, stereoptical data, and spectral changes). Also consider instrumental data, such as vertical changes, horizontal changes, and internal strains and tilt, including time history.
	Correlate movements with groundwater, weather, vibration, and human activity.

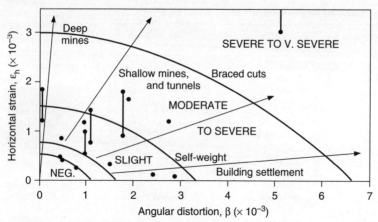

Figure 6.1 Relationship of damage to angular distortion and horizontal extension strain. (*From Boscardin and Cording, 1989, reprinted with permission from the American Society of Civil Engineers.*)

trench excavations (Sec. 6.9) and dam failures (Sec. 6.10). These types of lateral movement that cause damage to structures are also included in this chapter.

6.2   Allowable Lateral Movement of Buildings

As compared to the settlement of buildings, there is less work available on the allowable lateral movement. To evaluate the lateral movement of buildings, a useful parameter is the horizontal strain ε_h, defined as the change in length divided by the original length of the foundation. Figure 6.1 shows a correlation between horizontal strain ε_h and severity of damage (Boone, 1996; originally from Boscardin and Cording, 1989). Assuming a 6-m- (20-ft-) wide zone of the foundation subjected to lateral movement, Fig. 6.1 indicates that a building can be damaged by as little as 3 mm (0.1 in.) of lateral movement. Figure 6.1 also indicates that a lateral movement of 25 mm (1 in.) would cause "severe" to "very severe" building damage.

It should be mentioned that in Fig. 6.1, Boscardin and Cording (1989) used a "distortion factor" in their calculation of angular distortion ß for foundations subjected to settlement from mines, tunnels, and braced cuts. Because of this distortion factor, the angular distortion ß by Boscardin and Cording (1989) in Fig. 6.1 is different from the definition (δ/L) used in Chap. 4.

Figure 6.2 Damage due to lateral movement. [*From Day (1997a), reprinted with permission from the American Society of Civil Engineers.*]

The severity of building damage caused by lateral movement will depend on the tensile strength of the foundation. Those foundations that can not resist the tensile forces imposed by the slope movement will be the most severely damaged. For example, Figs. 6.2 and 6.3 show damage to a tilt-up building. For a tilt-up building, the exterior walls are cast in segments upon the concrete floor slab and then once they gain sufficient strength, they are tilted up into position. The severe damage shown in Figs. 6.2 and 6.3 was caused by slope movement, which affected the tilt-up building because it was constructed near the top of the slope. Figure 6.2 shows lateral separation of the concrete floor slab at the location of a floor joint. Figure 6.3 shows separation at the junction of two tilt-up panels. Because of the presence of joints between tilt-up panels and joints in the concrete floor slab, the building was especially susceptible to slope movement, which literally pulled apart the tilt-up building.

Those foundations that have joints or planes of weakness, such as the tilt-up building shown in Figs. 6.2 and 6.3, will be most susceptible to damage from lateral movement. Buildings having a mat foundation or a posttensioned slab would be less susceptible to damage because of the high tensile resistance of these foundations.

Figure 6.3 Joint separation between wall panels. [*From Day (1997a), reprinted with permission from the American Society of Civil Engineers.*]

6.3 Rock Fall

A rock fall is defined as a relatively free-falling rock or rocks that have detached themselves from a cliff, steep slope, cave, arch, or tunnel (Stokes and Varnes, 1955). The movement may be by the process of a vertical fall, by a series of bounces, or by rolling down the slope face. The free-fall nature of the rocks and the lack of movement along a well-defined slip surface differentiate a rock fall from a rockslide.

A rock slope or the rock exposed in a tunnel is characterized by a heterogeneous and discontinuous medium of solid rocks that are separated by discontinuities. The rocks composing a rock fall tend to detach themselves from these preexisting discontinuities in the slope or tunnel walls. The sizes of the individual rocks in a rock fall are governed by the attitude, geometry, and spatial distribution of the rock discontinuities. The basic factors governing the potential for a rock fall include: (1) the geometry of the slope or tunnel, (2) the system of joints and other discontinuities and the relation of these systems to possible failure surfaces, (3) the shear strength of the joints and discontinuities, and (4) destabilizing forces such as water pressure in the joints, freezing water, or vibrations (Piteau and Peckover, 1978).

For slopes, the main measures to prevent damage due to a rock fall are to alter the slope configuration, retain the rocks on the slope, intercept the falling rocks before they reach the structure, or direct the falling rocks around the structure. Altering the slope can include such measures as removing the unstable or potentially unstable rocks, flattening the slope, or incorporating benches into the slope (Piteau and Peckover, 1978). Measures to retain the rocks on the slope face include the use of anchoring systems (such as bolts, rods, or dowels), shotcrete applied to the rock slope face, or retaining walls. Intercepting or deflecting falling rocks around the structure can be accomplished by using toe-of-slope ditches, wire mesh catch fences, and catch walls (Peckover, 1975). Recommendations for the width and depth of toe-of-slope ditches have been presented by Ritchie (1963) and Piteau and Peckover (1978).

Of all engineering structures, tunnels are most vulnerable to small troublesome and large catastrophic failures (Feld and Carper, 1997). One reason is because the tunnel excavation allows for the sudden release of large confining pressures, resulting in rock strains which are not easily predicted. Another reason is the unpredictability of subsurface conditions, such as the possibility of encountering groundwater which can flood the tunnel. There are numerous case studies of the collapse of tunnels excavated in rock (for example, see Sec. 3.2.2 of Feld and Carper, 1997). Because of the numerous tunnel disasters, there has been the development of tunnel coring or boring machines (for soft and hard rock) that can provide some protection during excavation of the heading. Other measures to help stabilize tunnel rock faces include anchoring systems (such as bolts, rods, or dowels) and shotcrete. Chemical grouts have also been used to fill in rock discontinuities and joints in order to stabilize tunnel walls. For example, Feld and Carper (1997) describe a major rehabilitation project in 1985 that used foamed chemical grouts to stabilize the deteriorating tunnel walls of the 81-year-old Mt. Washington tunnel in Pittsburgh.

Figure 6.4 Rock fall (arrow indicates impact with house).

6.3.1 Case studies

The following case studies deal with rock fall in slopes. These case studies show that the behavior of a soft sedimentary rock fall can be much different from a rock fall in hard rock.

Case study 1. The rock fall occurred during the night when, according to the National Weather Service, about 4 cm (1.5 in.) of rainfall was recorded in the area. The tenant, asleep in the house, was awakened during the rainstorm by a loud noise and vibration of the house. Figures 6.4 and 6.5 show the rock fall debris in the rear yard and the impact damage to the house.

It was not clear if one or more rocks fell from the slope; however, for the purpose of this discussion, the rock fall will be referred to as a singular rock. During the night, the rock dislodged itself from the slope at the location shown in Fig. 6.6, fell into the rear yard, and then a portion impacted the rear of the house. The estimated weight of the rock that fell from the slope was on the order of 3.6 to 5.4 Mg (4 to 6 tons). Figure 6.7 shows a cross section through the house and rear yard slope.

The cause of the rock fall was an oversteepened cut slope with adversely oriented joint planes. Water (from the rainstorm) seeping

Figure 6.5   Close-up of impact with house.

along the joint planes provided the additional pressure and softening that was needed to dislodge the rock.

The classification of soil and sedimentary rock is based on grain size. Geologists and soil engineers frequently use different classification systems. Table 6.2 summarizes the grain size distribution from a sample of the rock fall. As shown in Table 6.2, the two systems differ the most in terms of the percent of sand and silt size particles. Geologists would probably classify the rock as a very fine sandstone to coarse siltstone. Because the rock contains 70 percent silt size particles (per the USCS), geotechnical engineers would tend to refer to the rock as a siltstone. It is rather academic as to what is the type of sedimentary rock, and thus it will be referred to as sandstone.

At this site, the sandstone was soft and friable. The dry density of the sandstone was 1.54 Mg/m³ (96 pcf). When a specimen of the on-site sandstone was placed in water, it completely disintegrated in less than 60 seconds. It was thus observed that the on-site sandstone was weakly cemented and tended to disintegrate when submerged in water. At the base of the rear yard slope (Fig. 6.7), there is a large amount of loose debris. The debris probably accumulated from previous rock falls and/or the process of slope weathering.

Given the size and weight of the rock fall and the very close proximity of the slope to the house, the amount of damage was relatively

Figure 6.6 Source of rock fall (asterisk shows source location).

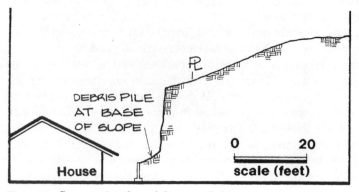

Figure 6.7 Cross section through house and slope.

TABLE 6.2   Grain Size Distribution of the Rock Fall

Size range or % particles	Grain-size scale used by American geologists (modified Wentworth scale)			Grain-size scale used by American engineers (Unified Soil Classification System, USCS)		
	Sand	Silt	Clay	Sand	Silt	Clay
(1)	(2)	(3)	(4)	(5)	(6)	(7)
Size range	2.0–0.062 mm	0.0062–0.004 mm	<0.004 mm	4.75–0.075 mm	0.075–0.002 mm	<0.002 mm
% particles by dry weight	40	53	7	25	70	5

minor (Fig. 6.5). This is because the rock was soft and broke apart upon impact with the rear yard concrete patio (Fig. 6.4). Instead of disintegrating upon impact, a hard and resistant rock would probably have ricocheted off the concrete and caused significantly more damage.

Case study 2.   The second case study involves two separate rock falls that occurred around November 1989 and early 1990. Figures 6.8 and 6.9 show photographs taken shortly after the second rock fall. In both photographs, the rock fall debris is indicated with an arrow. The unstable portion of the slope is near the top, where blocks of rock toppled off and then tumbled down the slope face. Figure 6.10 shows a cross section through the slope. Damage to the area at the toe of slope was slight and consisted mainly of minor damage to trees, parked cars, and the rear wall of a toe-of-slope building.

The cause of the two separate rock falls was a near vertical cut slope near the top of slope and adversely oriented joint planes. Figure 6.8 shows that one series of joint planes is near vertical. The Rose Canyon Fault zone is only about 120 to 150 m (400 to 500 ft) from the site (Kennedy, 1975) and movement on this nearby fault may have contributed to the fracturing of the rock. Since both rock falls occurred in the winter rainy season, water (from rainstorms) seeping along the joint planes may have provided additional water pressure and softening that dislodged the rock.

The rock is predominantly a siltstone. Grain size distribution analysis (ASTM D 422-90; ASTM, 1997e) of the rock indicate that most of the particles are from 0.004 to 0.06 mm in size. Samples of the rock were also tested in the direct shear apparatus, and the results indicate a cohesion of about 12 kPa (250 psf). This is a low

Figure 6.8 Location of rock fall (asterisk shows source of rock fall; arrow points to rock fall debris at toe of slope).

value and indicates that the siltstone is poorly cemented. Field observations also revealed that the siltstone was soft and friable. The average dry density of the siltstone was 1.44 Mg/m^3 (89.6 pcf) with a standard deviation of 0.04 Mg/m^3 (2.6 pcf).

Given the size of the rock fall and the close proximity of toe-of-slope structures, the amount of damage was again relatively minor. This is because the rock was weakly cemented and broke apart as it tumbled down the slope face and impacted with the toe of slope. Once again, a hard and resistant rock would have caused significantly more damage than the on-site rock, which disintegrated upon impact.

Case study 3. The third case study involves a rock fall in hard rock. In this case, a large rock detached itself from the slope and landed on

Figure 6.9 View from top of slope (arrow points to rock fall debris at toe of slope).

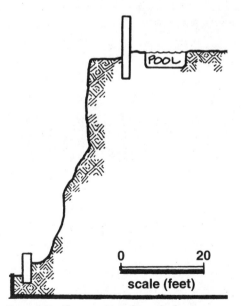

Figure 6.10 Cross section through slope.

Figure 6.11   Rock fall in Santiago Peak volcanics.

Figure 6.12   Source location of rock fall.

the roof of a house. The house is a one-story structure, having typical wood-frame construction and a stucco exterior. Figure 6.11 shows a view of the fallen rock. Figure 6.12 shows the location where the large rock detached itself from the slope.

The rock that impacted the house was part of the Santiago Peak volcanics, which consists of an elongated belt of mildly metamorphosed volcanic rock. The rock is predominately dacite and andesite.

Figure 6.13 Damage to kitchen area.

The Santiago Peak volcanics are hard and extremely resistant to weathering and erosion (Kennedy, 1975).

The part of the house below the area of impact of the rock was severely damaged as shown in Figs. 6.13 and 6.14. The oven, refrigerator, and other items are turned over at an angle in Fig. 6.13. The hanging baskets in the upper left corner of Fig. 6.13 provide a vertical plane of reference. Figure 6.14 shows the collapse of walls in the adjacent bathroom. The damage shown in Figs. 6.13 and 6.14 was due to the force of impact which crushed the roof and interior walls that were beneath the rock.

There was also damage in the house at other locations away from the area of rock impact. For example, Fig. 6.15 shows interior wallboard cracking and distortion of the door frames. Note that the door frames in Fig. 6.15 were originally rectangular, but are now highly distorted. Figure 6.16 shows how the roof literally opened up from the impact of the rock. Figure 6.17 shows damage to a roof beam in the house.

For this case, the rock remained intact during its detachment from the slope and impact with the house. Damage was severe because the house absorbed the entire concentrated energy of the rock fall. This is to be contrasted with the rock fall for cases 1 and 2, where the soft rock broke apart as it fell down the slope and then disintegrated upon impact with the toe of slope.

Figure 6.14 Damage to bathroom.

Figure 6.15 Damage to interior wallboard and distortion of door frames.

Figure 6.16 Damage to the roof.

Figure 6.17 Damage to roof beam.

Summary. In summary, the three case studies show examples of damage that can be caused by a rock fall. A factor in the amount of damage was the rock hardness, where the soft rock tended to break apart upon impact with the slope or toe of slope. In contrast, the rock fall in the Santiago Peak volcanics (hard rock) caused severe damage as shown in Figs. 6.13 to 6.17. The tools used by the forensic expert to investigate rock falls should include geologic mapping and research, subsurface exploration, laboratory testing, and engineering and geologic analyses to identify factors (such as adverse joint systems, seepage pressures, etc.) that contributed to the rock fall.

6.4   Surficial Slope Failures

Surficial failures of slopes are quite common throughout the United States (Day and Axten, 1989; Wu et al., 1993). In southern California, surficial failures usually occur during the winter rainy season, after a prolonged rainfall or during a heavy rainstorm, and are estimated to account for more than 95 percent of the problems associated with slope movement upon developed properties (Gill, 1967). Figure 6.18 illustrates a typical surficial slope failure. The surficial failure by definition is shallow with the failure surface usually at a depth of 1.2 m (4 ft) or less (Evans, 1972). In many cases, the failure surface is parallel to the slope face. The common surficial failure mechanism for clay slopes in southern California is as follows:

1. During the hot and dry summer period, the slope face can become desiccated and shrunken. The extent and depth of the shrinkage cracks depend on many factors, such as the temperature and humidity, the plasticity of the clay, and the extraction of moisture by plant roots.

2. When the winter rains occur, water percolates into the fissures, causing the slope surface to swell and saturate, with a corresponding reduction in shear strength. Initially, water percolates downward into the slope through desiccation cracks and in response to the suction pressures of the dried clay.

3. As the outer face of the slope swells and saturates, the permeability parallel to the slope face increases. With continued rainfall, seepage develops parallel to the slope face.

4. Because of a reduction in shear strength due to saturation and swell coupled with the condition of seepage parallel to the slope face, failure occurs.

Terraced Lots Above

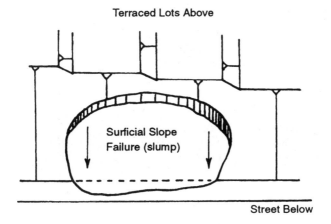

A) PLAN VIEW

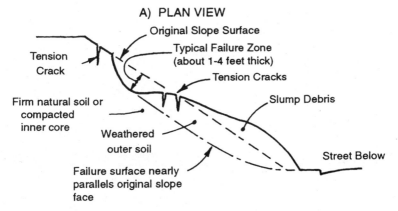

B) SECTION VIEW

Figure 6.18 Illustration of typical surficial slope failure. (*a*) Plan view; (*b*) cross-sectional view.

6.4.1 Stability analysis

A factor of safety is used to determine the stability of a slope. A factor of safety of 1.0 indicates a failure condition, while a factor of safety greater than 1.0 indicates that the slope is stable. The higher the factor of safety, the higher the stability of a slope.

To determine the factor of safety F for surficial stability, the following equation is commonly used:

$$F = \frac{c' + \gamma_b D \cos^2 \alpha \tan \phi'}{\gamma_t D \cos \alpha \sin \alpha} \tag{6.1}$$

This equation can be derived by assuming an infinite slope with seepage parallel to the slope face to a depth D (p. 356, Lambe and Whitman, 1969). Since steady-state flow conditions are assumed, effective shear strength soil parameters (ϕ' = effective friction angle; c' = effective cohesion) must be used in the analysis. In Eq. (6.1), α = slope inclination, γ_b = buoyant density of the soil, and γ_t = wet density of the soil. The parameter D is also the depth of the failure surface where the factor of safety is computed. The factor of safety for surficial stability [Eq. (6.1)] is highly dependent on the effective cohesion value of the soil. Automatically assuming c' = 0 for Eq. (6.1) may be overly conservative.

Because of the shallow nature of surficial failures, the normal stress on the failure surface is usually low. Studies have shown that the effective shear strength envelope for soil can be nonlinear at low effective stresses (Maksimovic, 1989). Because of the nonlinear nature of the shear strength envelope, the shear strength parameters obtained at high normal stresses can overestimate the shear strength of the soil and should not be used in Eq. (6.1) (Day, 1994c).

6.4.2 Surficial failures

Cut slope. Figure 6.19 shows a picture of a surficial slope failure in a cut slope. The slope is located in Poway, California, and was created in 1991 by cutting down the hillside during the construction of the adjacent road. The cut slope has an area of about 400 m² (4000 ft²), a maximum height of 6 m (20 ft), and a slope inclination that varies from 1.5:1 (34°) to 1:1 (45°). These are rather steep slope inclinations, but are not uncommon for cut slopes in rock. The failure mechanism is a series of thin surficial failures, about 0.15 m (0.5 ft) thick. The depth to length ratio (D/L) of a single failure mass is around 5 to 6 percent. Using this ratio, the slides shown in Fig. 6.19 fall within the classification range (3 to 6 percent) that Hansen (1984) has defined as "shallow surface slips."

The type of rock exposed in the cut slope is the Friars formation. This rock is of middle to late Eocene and has thick layers of nonmarine lagoonal sandstone and claystone (Kennedy, 1975). The clay minerals are montmorillonite and kaolinite. The Friars formation is common in San Diego and Poway, California, and is a frequent source of geotechnical problems such as landslides and heave of foundations.

The cause of the surficial slope failures shown in Fig. 6.19 was weathering of the Friars formation. Weathering breaks down the rock and reduces the effective shear strength of the material. The weathering process also opens up fissures and cracks which increases the per-

Figure 6.19 Surficial failure, cut slope for road.

meability of the near surface rock and promotes seepage of water parallel to the slope face. This weathering process is illustrated in Fig. 6.20 (Ortigao et al., 1997). Failure will eventually occur when the material has weathered to such a point that the effective cohesion approaches zero.

Fill slope.   Figure 6.21 shows a surficial slope failure in a fill slope. Studies by Pradel and Raad (1993) indicate that in southern California, fill slopes made of clayey or silty soils are more prone to develop the conditions for surficial instability than slopes made of sand or gravel. This is probably because water tends to migrate downward in sand or gravel slopes, rather than parallel to the slope face.

As shown in Fig. 6.21, surficial failures cause extensive damage to landscaping. Surficial failures can even carry large trees downslope

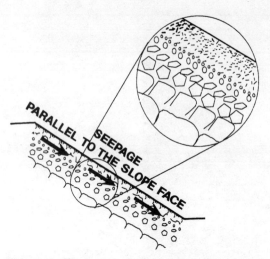

Figure 6.20 Illustration of near-surface weathering of claystone and zone of seepage parallel to slope face. (*Adapted from Ortigao et al., 1997, reprinted with permission from the American Society of Civil Engineers.*)

(see Fig. 6.21). Besides the landscaping, there can be damage to the irrigation and drainage lines. The surficial failure can also damage appurtenant structures, such as fences, walls, or patios.

A particularly dangerous condition occurs when the surficial failure mobilizes itself into a debris flow. In such cases, severe damage can occur to any structure located in the path of the debris flow. Figure 6.22 shows partial mobilization of the surficial failure, which flowed over the sidewalk and into the street. Surficial failures, such as the failure shown in Figs. 6.21 and 6.22, can be sudden and unexpected, without any warning of potential failure. Other surficial failures, especially in clays, may have characteristic signs of imminent failure. For example, Fig. 6.23 shows a clay slope having a series of nearly continuous semicircular ground cracks. After this picture was taken (during the rainy season), this slope failed in a surficial failure mode.

Surficial failures can also develop on the downstream face of earth dams. For example, Sherard et al. (1963) state:

> Shallow slides, most of which follow heavy rainstorms, do not as a rule extend into the embankment in a direction normal to the slope more than 4 or 5 feet [1.2 to 1.5 m]. Some take place soon after construction, while others occur after many years of reservoir operation.... Shallow surface slips involving only the upper few inches of the embankment have sometimes

Figure 6.21 Surficial slope failure in a fill slope.

occurred when the embankment slopes have been poorly compacted. This is a frequent difficulty in small, cheaply constructed dams, where the construction forces often do not make the determined effort necessary to prevent a loose condition in the outer slope. The outer few feet then soften during the first rainy season, and shallow slides result.

Effect of vegetation. A contributing factor in the instability of the slope shown in Fig. 6.23 was the loss of vegetation due to fire. Also notice in Figs. 6.19 and 6.21 that the surficial failures appear to have developed just beneath the bottom of the grass roots. Roots can provide a large resistance to shearing. The shear resistance of root-permeated homogeneous and stratified soil has been studied by Waldron

Figure 6.22 Partial mobilization of surficial failure.

(1977) and Merfield (1992). The increase in shear strength due to plant roots is due directly to mechanical reinforcement of the soil and indirectly by removal of soil moisture by transpiration. Even grass roots can provide an increase in shear resistance of the soil equivalent to an effective cohesion of 3 to 5 kPa (60 to 100 psf) (Day, 1993). When the vegetation is damaged or destroyed due to fire, the slope can be much more susceptible to surficial failure such as shown in Fig. 6.23.

6.5 Gross Slope Failures

Gross slope failures involve shear displacement of the entire slope. Terms such as *fill slope failures,* and *earth* or *rock slumps,* have been used to identify similar processes. The most common method of analysis for a gross slope failure is the "method of slices," where the failure mass is subdivided into vertical slices and the factor of safety is calculated based on force equilibrium equations. The objective of the slope stability analysis is to obtain the factor of safety. For those cases where a gross slope failure has occurred, the method of slices can be used to back-calculate the shear strength of the failed material by assuming a factor of safety of 1.0.

Figure 6.23 Ground cracks associated with incipient surficial instability.

The factor of safety for gross slope stability depends on the slope geometry, soil parameters such as the shear strength and wet density, and pore pressure conditions. Commonly used methods of slices to obtain the factor of safety are the Bishop method of slices (Bishop, 1955) and the Janbu method of slices (Janbu, 1957, 1968).

Duncan (1996) states that the nearly universal availability of computers and much improved understanding of the mechanics of slope stability have brought about considerable change in the computational aspects of slope stability analysis. Analyses can be done much more thoroughly, and, from the point of view of mechanics, more accurately than was possible previously. However, problems can develop because of a lack of understanding of soil mechanics, soil strength, and the computer programs themselves, as well as the inability to analyze the results in order to avoid mistakes and misuse (Duncan, 1996).

Foundation cracks may initially develop because of concrete shrinkage. The shrinkage cracks tend to act as planes of weakness in the foundation. The lateral slope movement will then cause these shrinkage cracks to open up. Damages become significant if the factor of safety of the slope approaches 1.0. In some cases, the gross slope movement may be to such an extent that the structure is so badly damaged that it has to be demolished

6.5.1 Case study

The purpose of this section is to describe the failure of Desert View Drive, located in La Jolla, California. The embankment supporting Desert View Drive started to fail in 1990 due to gross instability. The author was retained as an expert for one of the homeowners (the plaintiff) on Desert View Drive, who experienced property damage due to the road embankment failure. The plaintiff sued two main parties, the city of San Diego (which owned Desert View Drive) and an excavation group, which cut into the embankment supporting Desert View Drive and contributed to the failure. The main emphasis of this case study will be the suit between the plaintiff and the city of San Diego, because the author is more knowledgeable of this aspect of the lawsuit. The lawsuit between these two parties was settled out of court in 1993.

Other homeowners in the area of the embankment failure also sued both the city and the excavation group. These adjacent homeowners have also settled out of court.

Site history. The plaintiff's house is located on the east side of Mount Soledad in La Jolla. The building pad for the house was constructed as part of a large residential development. Mass-grading operations were used to cut down the high areas and fill in the lower canyons to create Desert View Drive and the adjacent building pads. No records or reports pertaining to the mass grading of the site could be found, but the mass grading was probably performed in late 1960 or early 1961.

On December 14, 1961, there was a major landslide adjacent to the site. Figure 6.24 shows two views of the landslide. The two photographs were taken from a helicopter. Figure 6.24a is a view toward the south and Fig. 6.24b is a view toward the west. Note in Fig. 6.24b

Figure 6.24a December 14, 1961, landslide, view to the south. (*Photograph courtesy San Diego Historical Society, photograph collection.*)

Figure 6.24b December 14, 1961, landslide, view to the west. (*Photograph courtesy San Diego Historical Society, photograph collection.*)

that, at the head of the landslide, the rear of one house is suspended in midair. The landslide damaged or destroyed nine houses that were under construction. The landslide was a roughly triangular mass of rock, about 180 m (600 ft) long, 80 m (250 ft) wide, and 25 m (80 ft) high. As shown in Fig. 6.24a, the plaintiff's property (site) was just outside the limits of the landslide.

The city of San Diego contracted for an independent investigation of the causes of the landslide. The investigation concluded that there were several factors responsible for the failure (Benton Engineering, 1962):

1. The rock bedding planes were inclined out of slope.

2. There was an intricate pattern of intersecting local faults and joints that provided planes of weakness on the sides of the landslide.

3. The construction of building pads in late 1960 or early 1961 resulted in a steepening of the slope.

4. Seasonal rains added weight and reduced the shear strength of the rock.

As part of the landslide investigation, the condition of the embankment fill supporting Desert View Drive was also analyzed. It was reported that the fill embankment had a relative compaction from about 80 to 86 percent, which was less than the city standard of a minimum relative compaction of 90 percent (Benton Engineering, 1962). Even lower relative compaction values of the embankment fill would have been obtained if a correction was made for oversize rock fragments (Day, 1989).

The plaintiff's building pad ("site" in Fig. 6.24a) lay undeveloped until about 1976, when the house was constructed. The house is a two-story, wood-frame structure having a concrete slab-on-grade. The floor area is about 170 m² (1800 ft²).

The first recorded problem with Desert View Drive embankment fill was a 1977 report (Alvarado Soils Engineering, 1977). In this report, it was stated:

> Our subsurface observations did not disclose any indication of excessive settlement and/or slope failures within the filled ground mass. However, separations between the road bed slabs and differential tilting of the gutter along Desert View Drive indicate that a certain amount of surface creep has taken place.

Another term for surface creep as described is *surficial instability.* Other than the report of surficial slope problems, there was no indica-

Figure 6.25 Cracking in Desert View Drive.

tion of gross instability until 1990, when an excavation group cut into
the fill embankment (across the street from the plaintiff's property) to
construct a building pad and house. Photographs taken of Desert
View Drive at the start of the excavation do not show any cracking or
separations in the roadway.

Desert View Drive embankment failure.   The two lines across Desert
View Drive in Fig. 6.24*a* indicate the approximate limits of the
embankment failure. The Desert View Drive embankment fill continued to move from 1990 through 1993. By adding up the widths
of ground cracks in the pavement and adjacent flatwork, the total
amount of horizontal movement from 1990 to 1993 was approximately 150 mm (6 in.). About half of this horizontal movement
occurred on the plaintiff's property. The most dramatic damages were
large separations and cracks in the concrete flatwork. There was also
damage to the house such as exterior stucco cracks, interior wallboard cracks, and horizontal displacement of the rear addition foundation.

Figure 6.25 (view to the north) shows the cracks in Desert View
Drive and the location of the plaintiff's property. The city periodically
patched the road surface. The arrow in Fig. 6.25 points to the utility
lines, which were installed above grade once the fill embankment

Note: Date of base reading is August, 1991

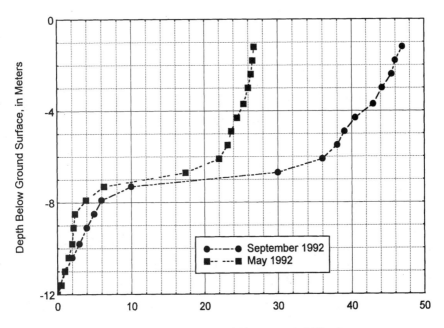

Figure 6.26   Inclinometer I-1.

started to move. The leakage of water from broken utility lines can increase the pore water pressure in a slope, which can accelerate the movement. The water billing records for all the homes involved in the embankment failure were reviewed, and they did not indicate a sudden surge in water usage. Also, borings and test pits excavated in the fill embankment did not encounter groundwater. A broken water line was not considered to be a contributing factor in the embankment failure.

Engineering analysis.   Figure 6.26 shows the horizontal (downslope) movement recorded by inclinometer I-1. Figure 6.24*a* shows the location of inclinometer I-1. The inclinometer was installed in August 1991, which was after the embankment had started to move. The primary zone of movement is about 7 to 8 m (23 to 26 ft) below ground surface, which is near the contact between the fill and underlying Ardath shale.

The factor of safety of the Desert View Drive embankment was determined by using cross section A-A' (Fig. 6.27). Cross section A-A'

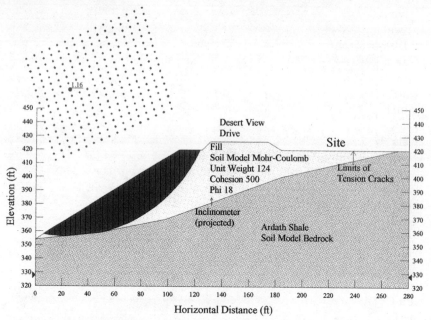

Figure 6.27 Cross section A-A'.

extends through the middle of the Desert View Drive embankment failure. For the slope stability analysis, the shear strength parameters were $\phi = 18°$ and $c = 24$ kPa (500 psf). The shear strength parameters were obtained from intact samples of the fill (Leighton and Associates, 1991). These shear values presented by another consultant were considered to be reasonable, and they were not verified by actual testing by the author. A total stress slope stability analysis was performed by using the SLOPE/W (Geo-Slope, 1991) computer program. The computer program, using the Janbu simplified method of slices, calculated a minimum factor of safety of 1.16 for the fill embankment. The location of the critical failure surface is shown in Fig. 6.27.

A factor of safety of 1.16 indicates marginal stability. When the excavation group cut into the Desert View Drive embankment in 1990, the factor of safety was lowered and movement began. The initial signs of distress were roadway cracks parallel to the utility backfill in the street. This would be consistent with the failure surface shown in Fig. 6.27.

Once the initial movement began, support was lost for the fill behind the failure zone. This unsupported fill also became unstable, and began to move. Thus the cracks progressed across Desert View

Drive and onto the plaintiff's property. Eventually, the majority of the fill wedge shown in Fig. 6.27 was moving horizontally (factor of safety of 1.0), as determined by inclinometer I-1, which indicated movement near the fill–Ardath shale contact. Thus, the failure of the Desert View Drive embankment was a progressive failure, starting as a circular failure (Fig. 6.27) and eventually increasing to a wedge-type failure of the entire fill embankment.

Basis for lawsuit. Several legal theories were developed as the basis for the lawsuit between the plaintiff and the city of San Diego. One argument was that the city, when it investigated the December 14, 1961, landslide, was aware that the Desert View Drive embankment had poorly compacted fill. The shear strength of poorly compacted fill is less than that of properly compacted (dense) fill. Compaction will densify the soil, and, because of the interlocking of soil particles, the friction angle generally increases. On the basis of experience with similar soils in this area and for a condition of a minimum of 90 percent relative compaction, total stress parameters are typically $c = 24$ kPa (500 psf) and $\phi = 28°$. Using these values and cross section A-A', the factor of safety of the embankment is 1.59, which is a condition of acceptable stability. It was thus argued that since the city of San Diego did not fix the defective condition (i.e., low compaction) of their embankment fill, they were partly responsible for its ultimate failure.

A second legal argument was the issue of subjacent support. The Desert View Drive embankment was providing lateral support for the plaintiff's property (Fig. 6.27). When the embankment fill moved horizontally, this lateral support was lost, resulting in damage.

Since the case settled out of court, it is not known if these legal theories would have prevailed in Superior Court.

Repair. To stop the movement of the Desert View Drive embankment, the city of San Diego installed 40 steel-reinforced concrete piers (see Fig. 6.28). The piers extended from ground surface and were anchored in the underlying Ardath shale. The reinforced-concrete piers were designed to increase the stability of the fill embankment by providing lateral resistance. The piers were located on the east side of Desert View Drive (Fig. 6.29), and most had a diameter of 1.4 m (4.5 ft) with a spacing of 2.4 m (8 ft) on center. Most piers were reinforced with 20 no. 18 bars and had a length of 23 m (76 ft).

Figure 6.28 (view to the east) shows the installation of the piers in October, 1993. The arrow in Fig. 6.28 points to a prefabricated steel cage. The site (plaintiff's house) is also identified in Fig. 6.28. The pier construction sequence started with the removal of the Desert

Figure 6.28 Construction of concrete piers at Desert View Drive.

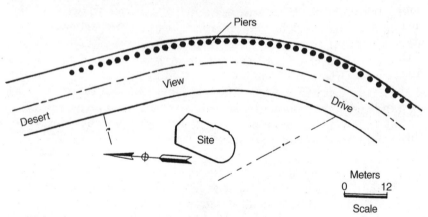

Figure 6.29 Plan view, location of piers.

View Drive sidewalk, followed by the excavation of the holes, then the prefabricated steel cages were inserted, and finally the holes were filled with concrete. The cost of the repair work was paid by the city of San Diego.

In summary, the failure of the Desert View Drive embankment was a progressive failure, starting as a circular gross instability, and eventually increasing to a wedge-type gross slope failure of the entire fill embankment. The tools used by the forensic engineer to investigate such failures include subsurface exploration, laboratory testing, field

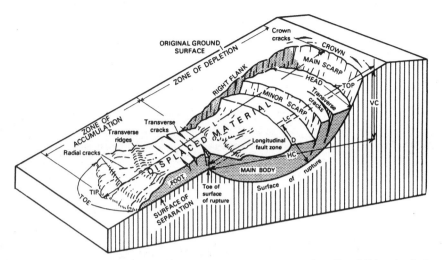

Figure 6.30 Landslide illustration. [*Reprinted with permission from Landslides: Analysis and Control, Special Report 176 (Varnes, 1978). Copyright 1978 by the National Academy of Sciences. Courtesy of the National Academy Press, Washington, D.C.*]

monitoring (inclinometers), and slope stability analysis. Two legal theories were used as the basis to sue the city of San Diego: (1) that city officials were aware of the defective condition (poor compaction) of the embankment, but did not fix this condition, and (2) the city failed to provide subjacent support to the plaintiff's property.

6.6 Landslides

The National Research Council states (1985):

> Landsliding in the United States causes at least $1 to $2 billion in economic losses and 25 to 50 deaths each year. Despite a growing geologic understanding of landslide processes and a rapidly developing engineering capability for landslide control, losses from landslides are continuing to increase. This is largely a consequence of residential and commercial development that continues to expand onto the steeply sloping terrain that is most prone to landsliding.

Figure 6.30 shows an example of a landslide and Table 6.3 presents common nomenclature used to describe landslide features (Varnes, 1978). The previous case study of the Desert View Drive embankment failure described the 1961 landslide that developed adjacent the site. Landslides are described as mass movement of soil or rock that involves shear displacement along one or several rupture surfaces, which are either visible or may be reasonably inferred (Varnes, 1978). As previously mentioned, it is the shear displacement along a distinct

TABLE 6.3   Common Landslide Nomenclature

Terms (1)	Definitions (2)
Main scarp	A steep surface on the undisturbed ground around the periphery of the slide, caused by the movement of slide material away from the undisturbed ground. The projection of the scarp surface under the displaced material becomes the surface of rupture.
Minor scarp	A steep surface on the displaced material produced by differential movements within the sliding mass.
Head	The upper parts of the slide material along the contact between the displaced material and the main scarp.
Top	The highest point of contact between the displaced material and the main scarp.
Toe, surface of rupture	The intersection (sometimes buried) between the lower part of the surface of rupture and the original ground surface.
Toe	The margin of displaced material most distant from the main scarp.
Tip	The point on the toe most distant from the top of the slide.
Foot	That portion of the displaced material that lies downslope from the toe of the surface of rupture.
Main body	That part of the displaced material that overlies the surface of rupture between the main scarp and toe of the surface of rupture.
Flank	The side of the landslide.
Crown	The material that is still in place, practically undisplaced and adjacent to the highest parts of the main scarp.
Original ground surface	The slope that existed before the movement which is being considered took place. If this is the surface of an older landslide, that fact should be stated.
Left and right	Compass directions are preferable in describing a slide, but if right and left are used they refer to the slide as viewed from the crown.
Surface of separation	The surface separating displaced material from stable material but not known to have been a surface on which failure occurred.
Displaced material	The material that has moved away from its original position on the slope. It may be a deformed or undeformed state.
Zone of depletion	The area within which the displaced material lies below the original ground surface.
Zone of accumulation	The area within which the displaced material lies above the original ground surface.

NOTE: Reprinted with permission from *Landslides: Analysis and Control, Special Report 176* (Varnes, 1978). Copyright 1978 by the National Academy of Sciences. Courtesy of the National Academy Press, Washington, D.C.

rupture surface that distinguishes landslides from other types of soil or rock movement such as falls, topples, or flows.

Landslides are generally classified as either rotational or translational. Rotational landslides are due to forces that cause a turning movement about a point above the center of gravity of the failure mass, which results in a curved or circular surface of rupture. Translational landslides occur on a more or less planar or gently undulatory surface of rupture. Translational landslides are frequently controlled by weak layers, such as faults, joints, or bedding planes; examples include the variations in shear strength between layers of tilted bedded deposits or the contact between firm bedrock and weathered overlying material.

Active landslides are those that are either currently moving or that are only temporarily suspended, which means that they are not moving at present but have moved within the last cycle of seasons (Varnes, 1978). Active landslides have fresh features, such as a main scarp, transverse ridges and cracks, and a distinct main body of movement. The fresh features of an active landslide enable the limits of movement to be easily recognized. Generally active landslides are not significantly modified by the processes of weathering or erosion.

Landslides that have long since stopped moving are typically modified by erosion and weathering, or may be covered with vegetation so that the evidence of movement is obscure. The main scarp and transverse cracks will have been eroded or filled in with debris. Such landslides are generally referred to as *ancient* or *fossil landslides* (Zaruba and Mencl, 1969; Day, 1995b). These landslides have commonly developed under different climatic conditions thousands or more years ago.

Many different conditions can trigger a landslide. Landslides can be triggered by an increase in shear stress or a reduction in shear strength. The following factors contribute to an increase in shear stress:

1. Removal of lateral support, such as erosion of the toe of the landslide by streams or rivers.

2. Application of a surcharge at the head of the landslide, such as the construction of a fill mass for a road.

3. Application of a lateral pressure, for example by the raising of the groundwater table.

4. Application of vibration forces, for example by an earthquake or construction activities.

Factors that result in a reduction in shear strength include:

■ Natural weathering of soil or rock

- Development of discontinuities, such as faults or bedding planes
- Increase in moisture content or pore water pressure of the slide plane material

6.6.1 Case study

The purpose of this case study is to illustrate the complex nature of forensic investigations of landslides. The case study deals with an ancient landslide located in San Diego, California. A portion of the ancient landslide became reactivated in 1994. Figure 6.31 shows a site plan of the portion of the ancient landslide that reactivated.

In the late 1970s, extensive geotechnical investigations were performed and the ancient landslide was discovered. The original development of the ancient landslide was due to translational movement probably more than 11,000 years ago (pre-Holocene epoch). Climatic studies have shown that the pre-Holocene climate was much wetter than the climate in southern California today. Annual rainfall in southern California during this time was probably 3 to 4 times greater than the present amount. The original development of the ancient landslide could have been due to the development of a high groundwater table or erosion which steepened or undermined the slope.

The ancient landslide was due to translational movement within the middle to late Eocene-aged Friars formation, which is characterized by gently dipping layers of sandstone or claystone (Kennedy, 1975). The claystone tends to be highly expansive and friable and usually resembles a stiff-fissured clay rather than a rock-type material. The typical spacing of the fractures within the stiff-fissured clay of the Friars formation is 2.5 cm to 5 cm (1 to 2 in.).

The actual rupture surface of the ancient landslide is a greenish remolded clay seam, which is polished with slickensides. The thickness of the clay seam is variable, but it is typically less than 0.3 m (1 ft). The depth of the rupture surface (clay seam) below ground surface varies up to a maximum of about 22 m (72 ft). Because of the presence of numerous internal shear surfaces, the ancient landslide was the result of translational movement progressing upslope, rather than a single massive failure event. Secondary landslides are also present along the head and flank of the ancient landslide.

Shear key. In order to increase the factor of safety of the ancient landslide, a shear key was proposed. A shear key is defined as a deep and wide trench cut through the ancient landslide mass and into intact bedrock below the slide. The trench is then backfilled with compacted fill to finish grade. A shear key is generally specified by mini-

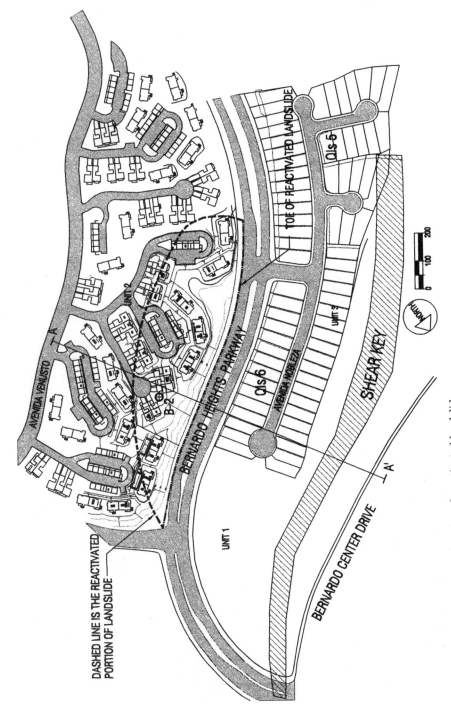

Figure 6.31 Site plan showing location of reactivated landslide.

DASHED LINE IS THE REACTIVATED
PORTION OF LANDSLIDE

TOE OF REACTIVATED LANDSLIDE

AVENIDA VENUSTO

UNIT 2

B-2

BERNARDO HEIGHTS PARKWAY

Qls-6

Qls-6

AVENIDA NOBLEZA

UNIT 3

SHEAR KEY

UNIT 1

BERNARDO CENTER DRIVE

A

A'

NORTH

0   100   200

175

mum key width and depth and a maximum back-cut angle. A shear key also normally contains a drainage system.

A shear key is excavated from ground surface to below the basal rupture surface, and then backfilled with soil that provides a higher shear strength than the original rupture surface. By interrupting the original weak rupture surface with a higher-strength soil, the factor of safety of the ancient landslide is increased. It is generally recognized that during the excavation of a shear key, there is a risk of reactivating the ancient landslide. To reduce this risk, a shear key is usually constructed in several sections with only a portion of the slide plane exposed at any given time.

Figure 6.31 shows the location of the shear key and Fig. 6.32 presents cross section A-A′, which shows how the shear key was designed to interrupt the basal rupture surface of the ancient landslide. The width of the shear key was determined from a rudimentary block-type failure analysis. In this case, the majority of the ancient landslide was represented by one main inclined block with the shear key being represented by another block. Forces between the blocks were ignored and the width of the shear key was determined as that width needed to increase the factor of safety of the blocks to 1.5. Because groundwater had not been encountered during the original geotechnical investigations, a groundwater table was not included in the block analysis. Also in the block analysis, the ancient landslide rupture surface was represented by cohesion = 2 kPa (40 psf), friction angle = 8°, and wet density (γ_t) of the landslide material = 2.0 Mg/m^3 (127 pcf).

The shear key was constructed in 1979 to 1980 during the mass grading of the site in order to create level building pads (see Fig. 6.31).

Multiunit residential structures, garages, and adjacent streets were built in several phases at the site. The first phase of construction took place in the early 1980s and the last phase was completed in 1994.

Reactivation of the ancient landslide. In 1994, a portion of the ancient landslide was reactivated. Figures 6.31 and 6.32 show the location of the reactivated portion. The landslide movement caused the formation of ground tension cracks and damage to homes and hardscape. Damage to the houses consisted primarily of interior and exterior wall cracking. The foundation systems for the homes were posttensioned slabs-on-grade, which helped mitigate damage.

The most significant damage was observed at the head and toe of the reactivated landslide. At the head of the landslide, there were tension-type damages due to the pulling apart of structures (Figs. 6.33 and 6.34), while at the toe of the landslide, there were buckling-type damages due to the compression of structures (Figs. 6.35 and 6.36).

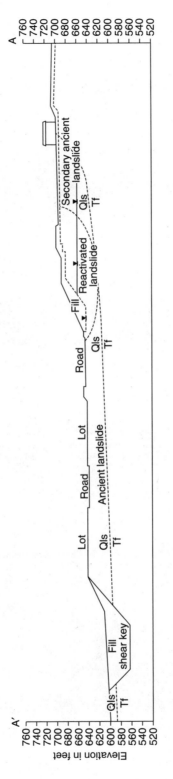

Figure 6.32 Cross section A-A'.

177

Figure 6.33 Head of the reactivated landslide (left and right sides of curb were initially level).

Figure 6.34 Damage to hardscape at the head of reactivated landslide.

Figure 6.35 Toe of reactivated landslide (arrows point to curb offset and buckling of asphalt).

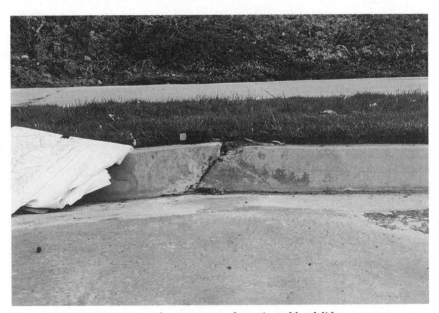

Figure 6.36 Compression-type damage at toe of reactivated landslide.

Inclinometer monitoring.   Inclinometers were installed in 1995. Figure 6.37 presents a plot of lateral displacement in the A and B direction versus depth below ground surface of inclinometer 2, which is located near the center of the landslide mass and shows the typical movement versus time. Figure 6.31 shows the location of this inclinometer, which was installed in boring B-2.

Figure 6.37 shows that the resultant displacement of the reactivated portion of the ancient landslide, as recorded by inclinometer 2 from June 1995 to December 1997, is about 3 cm (1.2 in.). The direction of the resultant vector of movement was generally downslope.

The total lateral movement of the reactivated portion of the landslide is greater than 3 cm (1.2 in.) because inclinometer 2 was installed after the movement started. The total lateral movement, based on the observed width of soil and hardscape cracks and separations, is probably in the range of 15 to 20 cm (6 to 8 in.).

Note in Fig. 6.37 that the movement occurred on a distinct failure plane, located about 22 m (72 ft) below ground surface. From the inclinometer data, the location of the slip surface for the reactivated portion of the ancient landslide is shown in Fig. 6.32.

Shear strength of slide plane.   The drained residual shear strength of the slide plane was determined by using a modified Bromhead ring shear apparatus (Stark and Eid, 1994). Back calculations of landslide shear strength indicate that the residual shear strength from ring shear tests is reasonably representative of the slip surface (Watry and Ehlig, 1995). The ring shear specimen is annular with an inside diameter of 7 cm (2.8 in.) and an outside diameter of 10 cm (4 in.). Drainage is provided by annular bronze porous stones secured to the bottom of the specimen container and the loading platen.

Remolded specimens were used for the ring shear and index property testing. The remolded specimens were obtained by air drying the slide plane clay seam, crushing it with a mortar and pestle, ball milling the crushed material, and processing it through the U.S. standard sieve no. 200. Distilled water was added to the processed soil until a water content approximately equal to the liquid limit was obtained. The specimen was then allowed to rehydrate for 7 days in a moist room. A spatula was used to place the remolded soil paste into the annular specimen container.

To measure the drained residual shear strength, the ring shear specimen was consolidated to 700 kPa (14,600 psf). The specimen was then unloaded to 50 kPa (1040 psf) and presheared by slowly rotating the ring shear base for one complete revolution using the hand wheel. After preshearing, the specimen was sheared at a drained displacement rate of 0.018 mm/min (0.0007 in./min). This is the slowest shear

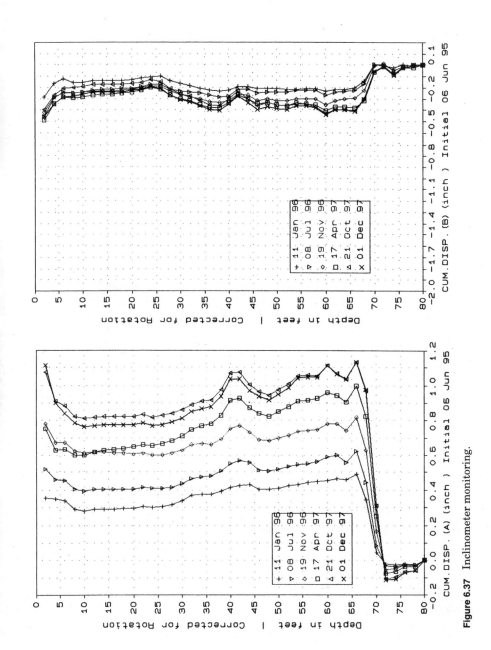

Figure 6.37 Inclinometer monitoring.

TABLE 6.4   Summary of Index Properties

Sample location (1)	Depth (2)	Sample type (3)	Clay size fraction (% <0.002 mm) (4)	Liquid limit (5)	Plastic limit (6)	Plasticity index (7)	USCS (8)
B-1	8.2–8.5 m (27–28 ft)	Disturbed (slide plane)	81	118	36	82	CH

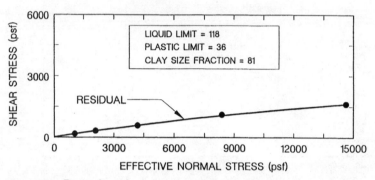

Figure 6.38 Drained residual shear strength envelope from ring shear test on slide plane material.

displacement rate possible and it has been successfully used to test soils that are more plastic than the slide plane materials.

After a drained residual strength condition was established at 50 kPa (1040 psf), shearing was stopped and the normal stress was increased to 100 kPa (2090 psf). After consolidation at 100 kPa (2090 psf), the specimen was sheared again until a drained residual condition was obtained. This procedure was also repeated for effective normal stresses of 200 kPa (4180 psf), 400 kPa (8350 psf), and 700 kPa (14,600 psf). The drained displacement rate of 0.018 mm/min (0.0007 in./min) was used for all stages of the multistage test.

Table 6.4 presents the index properties for the slide plane material. The slide plane material has a liquid limit that exceeds 100 and classifies as a highly plastic clay according to the Unified Soil Classification System. The liquid limit is high because the Friars formation consists primarily of montmorillonite clay particles (Kennedy, 1975).

Figure 6.38 presents the drained residual failure envelope and Fig. 6.39 shows the stress-displacement plots for the ring shear test on the slide plane sample. It can be seen in Fig. 6.38 that the failure envelope is nonlinear. If a secant failure envelope is assumed to pass through the origin and the shear stress at an effective normal stress

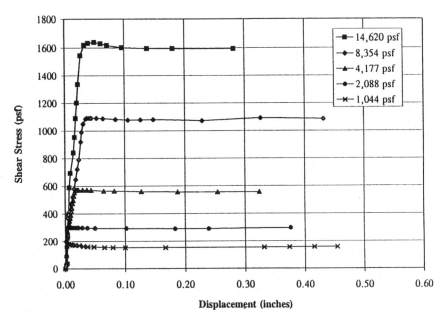

Figure 6.39 Shear stress versus displacement from ring shear test on slide plane material.

of 100 kPa (2090 psf), the secant friction angle is 8.2°. If a secant failure envelope is assumed to pass through the origin and the shear stress at an effective normal stress of 700 kPa (14,600 psf), the secant friction angle is 6.2°.

Stability analysis. The stability analysis of the reactivated portion of the ancient landslide was performed by using an effective stress analysis. The actual calculations were performed by the SLOPE/W (Geo-Slope, 1991) slope stability computer program, which uses the Janbu method of slices. Figure 6.40 shows a blowup of Fig. 6.32 for that portion of the ancient landslide that became reactivated.

The groundwater table shown in Fig. 6.40 was determined from piezometers. In southern California, it is not unusual for a groundwater table to develop after a site is developed. Once developed, the main sources of water infiltration into the ground are irrigation and leaky water pipes. Water due to irrigation is usually equivalent to 4 or 5 times the average annual rainfall [about 25 cm (10 in.)] at the site.

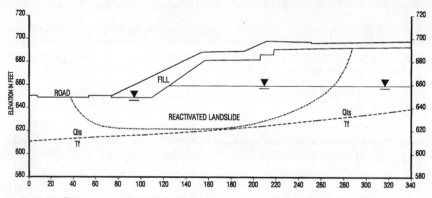

Figure 6.40 Cross section A-A′ showing reactivated portion of ancient landslide.

The model for the slope stability analysis was created using the following parameters:

1. The cross section shown in Fig. 6.40.
2. The actual nonlinear effective shear strength envelope of the slide plane material (Fig. 6.38).
3. The location of the slip surface as shown in Fig. 6.40, which was determined from the inclinometer data.
4. The location of the groundwater table, which was obtained from piezometer readings.
5. Wet density $\gamma_t = 2.0$ Mg/m³ (127 pcf) for the Friars formation.
6. Failure at the toe of the reactivated portion of the ancient landslide was assumed to be in fractured Friars formation.

Based on the input parameters, the factor of safety as calculated by the SLOPE/W (Geo-Slope, 1991) stability program was 0.92 for cross section A-A′. The factor of safety was also calculated for other cross sections through the flanks of the reactivated portion of the ancient landslide, and the weighted factor of safety for the entire reactivated landslide is about 1.0, which is consistent with the landslide movement at the site.

Destabilizing effects. The shear key was designed and constructed during the original development of the site. The shear key design addressed stabilization of the large-scale ancient landslide depicted on Fig. 6.32, but no analysis was conducted to determine the probability of intermediate failure within the slide mass due to the proposed construction at the site.

The site development created the conditions which were partly responsible for reactivation of a portion of the ancient landslide. The first destabilizing effect involves the excavation of the road cut for the main road (Bernardo Heights Parkway) located at the toe of that portion of the ancient landslide that reactivated. The road cut reduced the overburden above the slip surface, which facilitated the development of a rupture surface at the toe of the reactivated portion of the ancient landslide. A majority of the compression features observed along the toe of the reactivated portion of the ancient landslide coincide with the area of deepest road excavation. This is consistent with the stability analysis, which shows that the critical slip surface toe is at the roadway area.

A second destabilizing effect which contributed to the reactivation of a portion of the ancient landslide was the development of a groundwater table. As previously mentioned, a main source of water is from irrigation on site and at a housing complex further up the hillside. The presence of the groundwater table would introduce seepage forces within the landslide mass resulting in a decrease in the factor of safety.

Stabilization of landslide.   An attempt was made to reduce the landslide movement by lowering the groundwater table. Horizontal drains were considered, but since the elevation of the groundwater table and the slope toe were nearly the same, it was decided that a horizontal drain would not reduce the groundwater table enough to increase the landslide stability.

Rather than use horizontal drains, it was decided to investigate the feasibility of installing wells and then pumping water from the wells in order to lower the groundwater table. In early 1996, three vertical wells were installed into the landslide mass. The typical depths of the wells were 15 m (50 ft), and they had a diameter of 76 cm (30 in.). After installation of the wells, the pumping rates were monitored for a period of approximately 2 months. During this time, the pumping rates were very low (on the order of 5 gpm) and there was not a lowering of the groundwater table. It was concluded that the wells were ineffective because of the low hydraulic conductivity of the clayey landslide mass.

In order to stop the landslide, it was proposed in 1996 to increase the stability of the landslide by installing a retaining system utilizing tieback anchors. The retaining system has not yet been installed and the landslide continues to move at a rate of approximately 1 cm/year (0.4 in./year).

Summary. In summary, the forensic investigation indicated several causes for the reactivation of a portion of an ancient landslide, as follows:

1. *Shear key design.* In the analysis of the shear key, the ancient landslide was assumed to consist of one main inclined block. This is a common type of analysis, but it was not representative of the condition at the site. The original development of the ancient landslide was due to translational movement progressing upslope, rather than a single massive block failure event. The translational movement progressing upslope provided for interior shear surfaces along which a portion of the ancient landslide became reactivated.

2. *Groundwater table.* No groundwater table was observed during the original geotechnical investigation, and therefore it was not included in the analysis. The development of a groundwater table after construction of the site was probably due to additional moisture infiltration from ground surface irrigation.

3. *Nonlinear shear strength envelope.* The original design used slide plane shear strength values of $c = 2$ kPa (40 psf) and $\phi = 8°$. These values are similar to values obtained from the ring shear test at low effective stresses. For example, the secant effective friction angle is 8.2° at a normal stress of 100 kPa (2090 psf). But because the residual shear strength envelope is nonlinear (Fig. 6.38), at high normal stresses the shear strength parameters used in the block analysis ($c = 2$ kPa, $\phi = 8°$) are too high. The secant friction angle at 700 kPa (14,600 psf) is 6.2°. The road cut at the toe of the reactivated portion of the ancient landslide decreased the distance to the basal rupture surface and allowed the reactivated portion of the ancient landslide to toe up at this location.

For this landslide, a better design approach would have been to install a series of parallel shear keys. For example, a shear key could have been installed at the toe of the ancient landslide and then a second parallel shear key could have been constructed just above the road cut in order to stabilize the upper half of the ancient landslide.

This case study illustrates that landslides can be some of the most complex and challenging failures investigated by the forensic engineer. Especially for landslides affecting developed property, the cost of the investigation and repair can be high. As in gross slope failures, the tools used by the forensic engineer to investigate landslides include document research, inclinometer and piezometer monitoring, subsurface exploration, sophisticated laboratory testing, and extensive geologic and engineering analyses (such as computer slope stability analyses) in order to determine the causes of failure and develop repair options.

6.7 Debris Flow

Debris flows cause a tremendous amount of damage and loss of life throughout the world. An example is the loss of 6000 lives from the devastating flows that occurred in Leyte, Philippines, on November 5, 1991, due to deforestation and torrential rains from tropical storm Thelma. Because of continued population growth, deforestation, and poor land-development practices, it is expected that debris flows will increase in frequency and devastation. Debris flow is commonly defined as soil with entrained water and air that moves readily as a fluid on low slopes. As shown in Fig. 6.22, in many cases there is an initial surficial slope failure that transforms itself into a debris flow (Ellen and Fleming, 1987; Anderson and Sitar, 1995, 1996). Figure 6.41 shows two views of a debris flow: the source area of the debris flow and the debris flow forcing its way into a house.

Debris flow can include a wide variety of soil-particle sizes (including boulders) as well as logs, branches, tires, and automobiles. Other terms, such as *mud flow, debris slide, mud slide,* and *earth flow,* have been used to identify similar processes. While categorizing flows based on rate of movement or the percentage of clay particles may be important, the mechanisms of all these flows are essentially the same (Johnson and Rodine, 1984).

The historical method is one means of predicting debris-flow activity in a particular area. For example, as Johnson and Rodine (1984) indicated, many alluvial fans in southern California contain previous debris-flow deposits which in the future will likely again experience debris flow. However, using the historical method for predicting debris flow is not always reliable. For example, the residences of Los Altos Hills experienced an unexpected debris flow mobilization from a road fill after several days of intense rainfall (Johnson and Hampton, 1969). Using the historical method to predict debris flow is not always reliable, because the area can be changed, especially by society's activities.

Johnson and Rodine (1984) stated that a single parameter should not be used to predict either the potential or actual initiation of a debris flow. Several parameters appear to be of prime importance. Two such parameters, which numerous investigators have studied, are rainfall amount and rainfall intensity. For example, Neary and Swift (1987) stated that hourly rainfall intensity of 90 to 100 mm/h

Figure 6.41 Two views of a debris flow. Upper photograph shows the area of detachment, and the lower photograph shows how the flow forced its way into a house.

(3.5 to 4 in./h) was the key to triggering debris flows in the southern Appalachians. Other important factors include the type and thickness of soil in the source area, the steepness and length of the slope in the source area, the destruction of vegetation due to fire or logging, and other society-induced factors such as the cutting of roads.

There are generally three segments of a debris flow: the source area, main track, and depositional area (Baldwin et al., 1987). The *source area* is the region where a soil mass becomes detached and transforms itself into a debris flow. The *main track* is the path over which the debris flow descends the slope and increases in velocity depending on the slope steepness, obstructions, channel configuration, and the viscosity of the flowing mass. When the debris flow encounters a marked decrease in slope gradient and deposition begins, the segment is called the *depositional area*.

6.7.1 Case study

This section deals with a debris flow adjacent the Pauma Indian Reservation, in San Diego County, California. The debris flow occurred in January 1980, during heavy and intense winter rains. A review of aerial photographs indicates that alluvial fans are being built at the mouths of the canyons in this area due to past debris flows.

The debris flow in January 1980 hit a house, which resulted in a lawsuit being filed. The author was retained in June of 1990 as an expert for one of the cross-defendants, a lumber company. The case settled out of court in July, 1990.

The lumber company had cut down trees on the Pauma Indian Reservation and it was alleged that the construction of haul roads and the lack of tree cover contributed to the debris flow. These factors were observed to be contributing factors in the debris flow. For example, after the debris flow, deep erosional channels were discovered in the haul roads, indicating that at least a portion of the soil in the flow came from the road subgrade.

Figure 6.42 presents a topographic map of the area, showing the location of the house and the path of the debris flow. The debris flow traveled down a narrow canyon (top of Fig. 6.42) that had an estimated drainage basin of 0.8 km^2 (200 acres). The complete extent of the source area could not be determined because of restricted access. The source area and main track probably extended from an elevation of about 980 m (3200 ft) to 400 m (1300 ft) and the slope inclination of the canyon varied from about 34° to 20°. At an elevation of about 400 m (1300 ft), the slope inclination changes to about 7°, corresponding to the beginning of the depositional area.

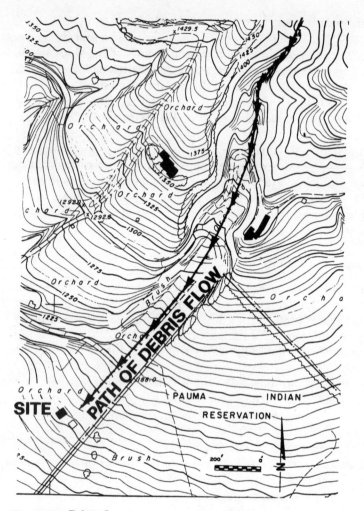

Figure 6.42 Debris flow.

Figure 6.43 presents a photograph of the house, which is a one-story single-family structure, having a typical wood-frame construction and a stucco exterior. It was observed in the area of the site that the debris flow was uniform with a thickness of about 0.6 m (2 ft). There was no damage to the structural frame of the house because of two factors: (1) The debris flow impacted the house and moved around the sides of the house, rather then through the house, because of the lack of any opening on the impact side of the house (Fig. 6.43). (2) The debris flow had traveled about 370 m (1200 ft) in the depositional area before striking the house and then proceeded only about 15 m

Figure 6.43 Photograph of site (arrows indicate height of the debris flow).

(50 ft) beyond the house. By the time the debris flow reached the house, its energy had been nearly spent.

A sample of the debris-flow material was tested for its grain-size distribution. The soil comprising the debris flow was classified as a nonplastic silty sand (SM), with 16 percent gravel, 69 percent sand, and 15 percent silt and clay. The material in the debris flow was derived from weathering of rock from the source area. Geologic maps of the area indicated that the watershed is composed of metamorphic rocks, probably the Julian schist.

On the basis of the historical method, for the site shown in Fig. 6.42, it seems probable that there will be future debris flows. Perhaps the best solution is to restrict the construction of houses in the depositional area. Notice in the center of Fig. 6.42 that two houses were built very close to the path of the debris flow, but they are situated at an elevation about 15 m (50 ft) above the depositional area and, hence, were unaffected by the debris flow.

In summary, two contributing factors in the debris flow were the cutting down of trees and the construction of dirt haul roads. The debris flow traveled a long way (370 m) in the depositional area before striking the house. Even if logging is restricted and haul roads eliminated, it is probable that there will be future debris flow. From an engineering standpoint, perhaps the best solution is to restrict the construction of houses in the alluvial fan depositional area.

6.8   Slope Softening and Creep

In many urban areas, there is a tendency toward small lot sizes because of the high cost of the land, with most of the lot occupied by building structures. A common soil engineering recommendation is to have the bottom edge of the footing at least 1.5 m (5 ft) horizontally from the face of any slope, irrespective of the slope height or soil type. This recommendation results in many buildings being constructed near the top of fill slopes.

Fill in slope areas is generally placed and compacted at near optimum moisture content, which is often well below saturation. After construction of the slope, additional moisture is introduced into the fill by irrigation, rainfall, groundwater sources, and leaking water pipes. At optimum moisture content, a compacted clay fill can have a high shear strength because of negative pore water pressures. As water infiltrates the clay, the slope softens as the pore spaces fill with water and the pore water pressures tend toward zero. If a groundwater table then develops, the pore water pressures will become positive. The elimination of negative pore water pressure results in a decrease in effective stress and deformation of the slope in order to mobilize the needed shear stress to maintain stability. This process of moisture infiltration into a compacted clay slope which results in slope deformation has been termed *slope softening* (Day and Axten, 1990).

Some indications of slope softening include the rear patio pulling away from the structure, pool decking pulling away from the coping, tilting of improvements near the top of the slope, stair-step cracking in walls perpendicular to the slope, and downward deformation of that part of the building near the top of the slope. In addition to the slope movement caused by the slope softening process, there can be additional movement due to the process of creep.

Creep is generally defined as an imperceptibly slow and more or less continuous downward and outward movement of slope-forming soil or rock (Stokes and Varnes, 1955). Creep can affect both the near-surface soil or deep-seated materials. The process of creep is frequently described as viscous shear that produces permanent deformations, but not failure as in landslide movement. Typically the amount of movement is governed by the following factors: shear strength of the clay, slope angle, slope height, elapsed time, moisture conditions, and thickness of the active creep zone (Lytton and Dyke, 1980).

6.8.1   Case study

This case study illustrates the typical procedures used by the forensic engineer to investigate the creep of slopes. The site is located near the top of Mt. Soledad in La Jolla, California. The original topography

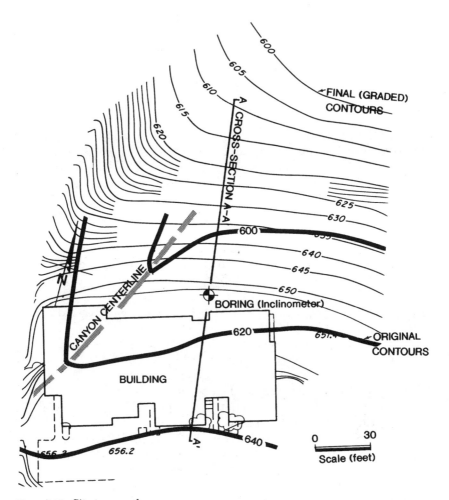

Figure 6.44 Site topography.

consisted of a canyon draining toward the northeast. In order to create a split-level building pad, the canyon was filled in with compacted clay. The grading occurred in June and July 1974. Figure 6.44 shows a topographic map of the site, with the original canyon elevations and the final split-level pad elevations.

During the mass grading in 1974, a fill slope was constructed at the rear of the building pad (see Fig. 6.44). The upper part of the slope has an inclination of 2:1 (horizontal:vertical). The slope inclination decreases at the base of the slope. The greatest depth of fill is at the top of the slope.

The original grading report lists the results of field density and moisture content tests performed during the grading of the site. This data indicates that the average moisture content, during placement of the fill, was 17.9 percent, with a standard deviation of 1.8 percent. Using the average dry density and moisture content of the fill, the average degree of saturation of the as-compacted fill was 80 percent.

Figure 6.44 also shows the location of the building constructed on the split-level building pad. The garage was constructed at the higher pad elevation [200 m (656 ft)], with the attached two-story living areas at the lower pad elevation [198 m (651 ft)]. The building was constructed in the late 1970s, and contains four condominium units. The building has a slab-on-grade foundation (living areas) and is of typical wood-frame construction. The exterior is covered with stucco and the interior walls are faced with wallboard.

Figure 6.45 shows the manometer survey, performed in June 1986, for the building shown in Fig. 6.44. The manometer survey indicated 55 mm (2.15 in.) of differential foundation movement. The low area from the manometer survey was at the rear of the building. There was also damage associated with the foundation movement. The damage tended to increase toward the rear of the building (top of fill slope). For example, Figs. 6.46 and 6.47 show typical cracking and separation of the rear-yard patio.

Based on the data obtained from the manometer survey, subsurface exploration was performed in January, 1987. A 150-mm- (6-in.-) diameter auger boring was excavated in the rear yard. Figure 6.44 shows the location of the auger boring. The boring log is shown in Fig. 6.48. The subsurface exploration revealed about 13 m (43 ft) of silty clay fill overlying the Ardath shale (bedrock).

After completion of the boring, an inclinometer casing was installed. In addition, soil samples recovered during the subsurface exploration were returned to the laboratory. The fill can be described as a silty clay of low plasticity (CL). The principal clay minerals are kaolinite and montmorillonite.

Table 6.5 presents a summary of moisture and density tests performed on intact samples obtained from the boring. Excluding the near-surface soil sample at a depth of 0.9 m (3 ft), the average degree of saturation of the fill in 1987 was 96 percent (Table 6.5). As previously mentioned, the average degree of saturation at the end of grading (July 1974) was 80 percent. This increase in moisture content and degree of saturation from 1974 to 1987 was probably due to the infiltration of moisture into the slope from rainfall and irrigation. Although no free water was observed during the subsurface exploration, the high degree of saturation could indicate localized zones of perched groundwater.

Drained direct shear tests were performed on samples of the silty clay fill, and the results indicate peak effective shear strength parameters of friction angle ($\phi' = 28°$) and cohesion ($c' = 2.4$ kPa, 50 psf).

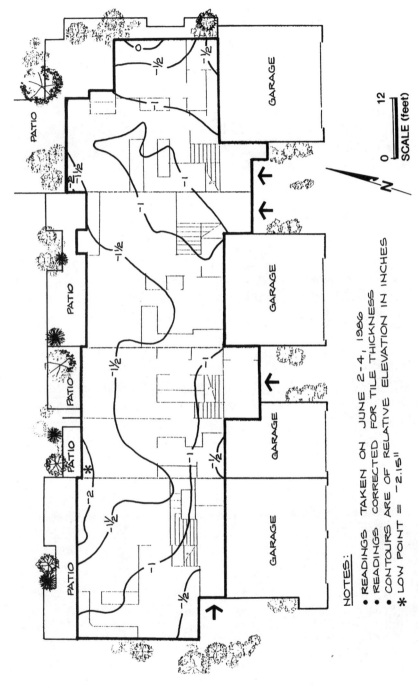

NOTES:
- READINGS TAKEN ON JUNE 2-4, 1986
- READINGS CORRECTED FOR TILE THICKNESS
- CONTOURS ARE OF RELATIVE ELEVATION IN INCHES
- ✳ LOW POINT = −2.15"

SCALE (feet)
0   12

GARAGE

GARAGE

GARAGE

GARAGE

PATIO

PATIO

PATIO

PATIO

PATIO

N

Figure 6.45 Manometer survey.

Figure 6.46 Damage to rear-yard patio.

Figure 6.47 Damage to rear-yard patio.

BORING LOG

Project/Client: CONDOMINIUMS, LA JOLLA, CALIFORNIA   F.N.:   N/A

Location: Rear yard   Date: 1-28-87
Estimated Surface Elevation (ft): 651   Total depth (ft): 50   Rig Type   6" Mini Flight Auger

Depth (Feet)	Sample Type	Sample Depth	Blow Count	Field Description By:
				Surface Conditions: At top of slope, behind building.
				Subsurface Conditions: FORMATION: Classification, color, moisture, tightness, etc.
0				FILL: From 0-43.0', Silty Clay, pale olive, moist, stiff; numerous fragments, medium green gray
	T	3		to pale yellow brown, claystone up to 3", mottled orange, concretion at 5'.
	T	9		
10				
	T	12		
	T	15		
	T	18		
20	T	21		
	T	24		At 24.0', Soil slightly darker and siltier, slightly fewer and irregular fragments.
				At 26.0', Very few fragments, soil slightly more moist.
30	T	29		
				At 34.0', Soil becomes wet.
				At 37.0', Soil becomes dark olive brown, wet, organic odor.
40	T	40		
				ARDATH SHALE: From 43.0'-45.0', Claystone, pale, green, damp, hard; fractured, varies to
				pale olive brown, stained pale orange.
50				Total Depth - 50.0'.
				No water.
				No caving.
60				

Notes: T = Shelby Tube

Figure 6.48 Boring log.

Figure 6.49 shows the results of the inclinometer monitoring. As previously mentioned, the inclinometer casing was installed in the auger boring (see Fig. 6.44). In Fig. 6.49, the vertical scale is depth below ground surface and the horizontal scale is the lateral (out-of-slope) movement.

The initial survey (base reading) was obtained on February 11, 1987. Additional successive surveys were performed from 1987 to 1995. As shown in Fig. 6.49, the inclinometer readings indicate a slow, continuous lateral movement (creep) of the slope. The inclinometer data does not show movement on a discrete failure surface, but rather indicates a progressive creep of the entire fill slope. The inclinometer recorded the greatest amount of lateral movement [33 mm

TABLE 6.5   Summary of Moisture and Density Tests

Sample location m (ft) (1)	Field dry density, Mg/m³ (pcf) (2)	Field moisture content, % (3)	Degree of saturation, % (4)
0.9 (3)	1.65 (103)	16.1	71
2.7 (9)	1.71 (107)	19.8	97
3.7 (12)	1.71 (107)	20.6	100
4.6 (15)	1.73 (108)	19.4	97
5.5 (18)	1.70 (106)	20.0	95
6.4 (21)	1.67 (104)	20.0	91
7.3 (24)	1.68 (105)	20.4	94
8.8 (29)	1.73 (108)	19.3	96
12.2 (40)	1.62 (101)	24.5	100

NOTE: Samples obtained from the auger boring (see Fig. 6.44 for location of the auger boring).

(1.3 in.)] at a depth of 0.3 m (1 ft). The amount of slope movement decreases with depth.

Figure 6.50 shows a plot of the amount of lateral slope movement versus time for readings at a depth of 0.9 m (1 ft) and 3.0 m (10 ft). This plot shows that the rate of slope movement is about 4.1 mm/year (0.16 in/year) at a depth of 0.9 m (1 ft), and about 1.3 mm/year (0.05 in/year) at a depth of 3.0 m (10 ft).

A slope stability analysis was performed by using the SLOPE/W (Geo-Slope, 1991) computer program. Using the slope configuration, soil shear strength parameters, wet density of the soil, and the pore water conditions as input, the computer program calculates the factor of safety of the slope by the Janbu simplified method of slices.

Figure 6.51 shows the slope configuration used in the stability analysis. Figure 6.44 shows the location of cross section A-A'. The cross section was developed from the pre- and postgrading topographic maps and the results of the boring (Fig. 6.48). The fill parameters used for the computer analysis were $\phi' = 28°$, $c' = 2.4$ kPa (50 psf), and $\gamma_t = 2.0$ Mg/m³ (127 pcf). The pore water pressures were assumed to be equal to zero.

According to the SLOPE/W (Geo-Slope, 1991), computer program, the minimum factor of safety of the fill slope is 1.28. Figure 6.51 also presents the results of the slope stability analysis. This figure shows the slip surface having the lowest factor of safety. The stability analysis indicates that the fill closest to the slope face has the lowest factor

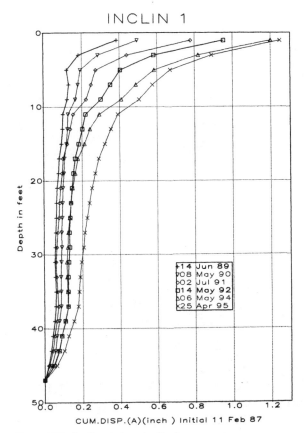

Figure 6.49 Inclinometer monitoring.

of safety, which is consistent with the greatest amount of movement recorded from the inclinometer.

The results of the subsurface exploration, laboratory testing, and stability analysis indicate that the causes of the fill slope movement are the following:

1. *Slope softening.* Slope softening is the process of moisture infiltration into the fill slope from rainfall and irrigation. The moisture infiltration into the slope causes an increase in wet density and a reduction in the negative pore water pressure as the fill proceeds from an unsaturated (S = 80 percent) to a saturated (S = 100 percent) or nearly saturated state. The increase in wet density and the decrease in effective stress of the clay slope results in deformation of the slope in order to mobilize the needed shear stress to maintain stability. On the basis of the high degree of sat-

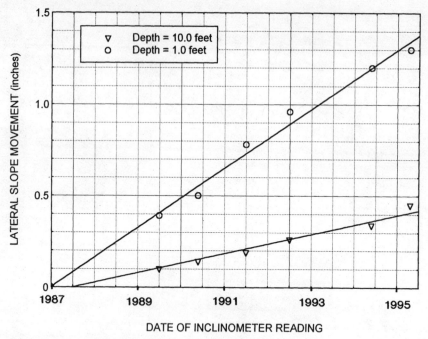

Figure 6.50   Lateral slope movement versus time.

uration (Table 6.5) from fill samples recovered during the subsurface exploration, most of the slope softening had occurred by
1987.

2. *Creep.*   There are apparently two factors contributing to the
creep of the slope:

   a. Loss of peak shear strength.   In the stability analysis, the
   peak effective shear strength parameters were used ($\phi' = 28°$,
   $c' = 2.4$ kPa, 50 psf). Figure 6.52 shows the shear stress versus horizontal deformation for the drained direct shear test
   specimen having a normal pressure during testing of 19 kPa
   (400 psf). The peak strength is identified in this figure. Note
   that with continued deformation, the shear strength decreases
   and reaches an ultimate value (which is less than the peak). It
   has been stated (Lambe and Whitman, 1969) that overconsolidated soil, such as compacted clay, has a peak drained shear
   strength which it loses with further strain, and thus the overconsolidated and normally consolidated strengths approach
   each other at large strains. This reduction in strength of the
   compacted clay as it deforms during drained loading can promote creep.

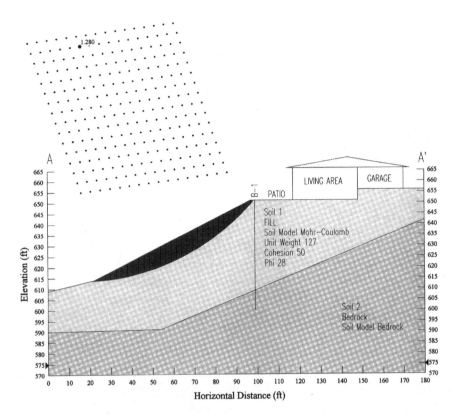

Figure 6.51 Cross section A-A'.

b. *Seasonal moisture changes.* It has been stated that seasonal moisture changes can cause creep of fill slopes. For example, Bromhead (1984) states:

> These strains reveal themselves (usually at the surface) as deformations or ground movement. Some movement, particularly close to the surface, occurs even in slopes considered stable, as a result of seasonal moisture content variation; creep rates of a centimeter or so per annum are easily reached. Because of the shallow nature it is only likely to affect poorly-founded structures.

At this site, there would be seasonal moisture changes that probably contributed to the near-surface creep of the fill slope.

No repairs were made to the building foundation. However, the damaged rear-yard patios and patio walls have been removed and replaced. Also, the drainage has been improved by regrading poorly draining areas and installing additional catch basins. The downspouts from the roof gutters have been connected to an underground drainage system.

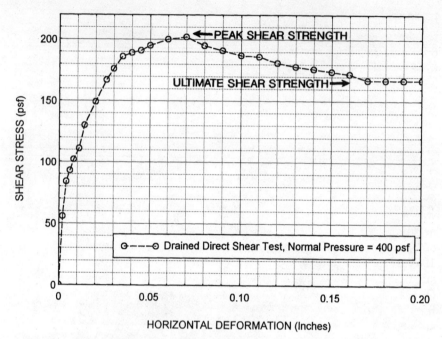

Figure 6.52 Drained direct shear test, silty clay fill.

This case study shows the difficulty in predicting the performance of compacted clay slopes. The stability of the slope depends on the ultimate moisture or groundwater conditions that may develop long after the compaction of the slope. The forensic engineer should determine what assumptions were made concerning the design of the slope and, in particular, the assumed design pore water pressure conditions.

In summary, this case study deals with fill slope deformation due to initial softening followed by creep. The initial compaction of the fill occurred in 1974 and the average degree of saturation during compaction was 80 percent. The subsurface exploration in 1987 revealed that the moisture content of the fill had increased and the average degree of saturation was 96 percent. This increase in moisture content was probably due to infiltration of water into the slope from rainfall and irrigation. The infiltration of moisture into the slope increased the wet density and decreased the effective stress of the compacted clay. This resulted in deformation of the slope in order to mobilize the needed shear stress to maintain stability, and this process is termed *slope softening.*

The manometer survey recorded 55 mm (2.15 in.) of slab-on-grade differential in 1986. The slab-on-grade tilted downward toward the top-of-slope (Fig. 6.45). In addition to the foundation differential, there was lateral movement and damage to the rear-yard patio (Figs. 6.46 and 6.47). An inclinometer was installed in 1987 and the monitoring from 1987 to 1995 shows there is not a discrete failure plane, but rather there is a progressive creep of the entire fill slope. Two factors affecting the creep of the slope are a loss in peak shear strength with further strain and near-surface seasonal moisture changes.

6.9 Trench Cave-Ins

A trench is generally defined as a narrow excavation made below the ground surface that may or may not be shored. Each year, numerous workers are killed or injured in trench or excavation cave-ins. These accidents typically are due to the fact that the trench was not shored or the shoring system was inadequate (Thompson and Tanenbaum, 1977). In the United States, fatalities in trench cave-ins represent a substantial portion of all construction fatalities. For example, the Ohio Bureau of Workers' Compensation indicated a total of 271 cave-ins during the period 1986–1990. The cause of death in many cases is suffocation. Other workers die from the force of the cave-in, which causes a crushing injury to the chest.

Petersen (1963) described the death of Peter Reimer, a construction worker, who decided that a trench needed no shoring because, he said, "It looks as solid as a brick wall." Reimer was alone in the trench when one side caved in, but he was only buried up to his chest. Three fellow workers started digging Reimer out, but they had to scramble to safety when the other side of the trench caved in. Reimer's body was uncovered half an hour later. As Petersen states, the first rule of a cave-in is to brace the excavation to prevent further cave-ins, then proceed with the rescue.

On large construction projects or deep excavations for buildings, money is generally available to perform an exploration program that defines both the soil and groundwater conditions over the full extent of the project. This results in an engineered shoring design that is commonly adjusted to satisfy the varying site conditions. However, utility-trench excavations are smaller projects with typically less money for subsurface exploration. Also, utility-trench excavations frequently extend over longer distances than large projects or deep building excavations. In many situations for utility-trench excavations, it is the contractor who determines which shoring system to install on the basis of the superintendent's experience and the observed soil and groundwater conditions. When trench walls are loose or sandy, the danger is apparent to the contractor and shoring is installed. As in the Reimer case (Petersen, 1963), many problems develop when the trench looks hard and compact, and the contractor will take chances and install minimal or no shoring on the assumption that it will be stable for a few hours.

A common form of shoring for unstable soil is close sheeting. Figure 6.53 presents a diagram of close sheeting ("Trench," 1984). The main components of the shoring system are continuous sheeting from the top to the bottom of the trench, which is held in place by stringers and cross braces (struts). In many cases, the cross braces are hydraulic jacks rather than wood members. Another common type of shoring uses a sliding trench metal shield (NAVFAC DM-7.2, 1982).

Many injuries occur after the utility lines have been placed in the bottom of the trench and it is time to backfill. If the trench has remained open for a long time, there can be much more pressure on the shoring system than when it was installed. A worker who is in the trench and knocks out a cross brace will release this built-up pressure and perhaps cause failure of other members or collapse. According to

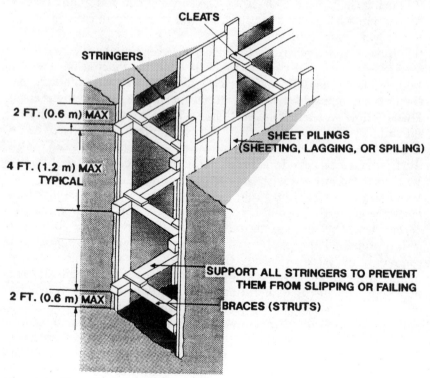

Figure 6.53 Close sheeting for running soil.

Petersen (1963), the following is the correct procedure to backfill a trench.

1. Backfill and compact the trench to a point just below the bottom cross brace.
2. Enter the trench and remove only the bottom cross brace.
3. Backfill and compact to the next level of cross braces.
4. Remove these cross braces and continue the procedure until only the uprights or sheeting remain in the ground.
5. Pull out the uprights or sheeting using either a backhoe or front-end loader.

When investigating trench cave-ins, the forensic engineer will need to investigate the type of shoring system and determine if the shoring system meets minimum state or local requirements. For example, the California Occupational Safety and Health Administration (Cal/OSHA) has developed a system to design shoring based on four different material types. The previous federal OSHA classification system ("Excavations," 1989) was to design shoring for either one of two soil conditions: hard compact soil or running soil. Running soil is defined as earth material where the angle of repose is approximately zero, as in the case of soil in a nearly liquid state or dry, unpacked sand that flows freely under slight pressure. Running material also includes loose or disturbed earth that can only be contained with close sheeting (*Standards,* 1991). Hard compact soil was defined as all earth material not classified as running soil.

The new Cal/OSHA shoring design system (effective September 25, 1991) categorizes soil and rock deposits in a hierarchy of stable rock, type A, type B, and type C, in decreasing order of stability (*Standards,* 1991). *Stable rock* is defined as solid rock that can be excavated with vertical sides and remain intact while exposed. The definitions of type A, B, and C materials are rather extensive, but in general, the classifications are based on the strength of the material. For example, if the trench were excavated in cohesive soil, then it would be type A if the unconfined compressive strength were greater than 150 kPa (1.5 tsf), type B if the unconfined compressive strength was between 50 and 150 kPa (0.5 and 1.5 tsf), and type C if the unconfined compressive strength was less than 50 kPa (0.5 tsf). California OSHA has tables (*Standards,* 1991) that provide the minimum timber shoring requirements for each soil type and a specific depth of trench. A type C soil would require a more substantial shoring system than a type A soil, given the same depth of the trench.

Besides the shoring system and soil type, other important factors to be considered by the forensic engineer during the investigation of a trench cave-in are as follows:

1. A temporary increase in the depth, length, or width of the trench.
2. The presence of vertical side slopes or steepened slopes which are not in conformance with the recommended trench configuration.
3. The excavation of a trench below the groundwater table.
4. The presence of surcharge loads from adjacent soil or material stockpiles, buildings, and other loads.
5. The presence of vibrations from passing traffic, earthquakes, jackhammers, and other sources.
6. The possibility that the trench was left open a long time.
7. A change in climate, such as frequent rainstorms or thawing of frozen ground.
8. Surface-water flow that was not diverted away from the top of the trench.

6.9.1 Case study

This section deals with a trench cave-in that occurred in downtown San Diego at the intersection of Fourth Avenue and Market Street on August 28, 1986, during the installation of a new sewer line. One worker was injured during the cave-in, which resulted in a lawsuit. For the purpose of this case study, the initiator of the lawsuit (plaintiff) will be referred to as the injured worker. The contractor excavating the trench will be referred to as either the contractor or the employer.

The author was retained in November 1988 as an expert for one of the cross-defendants, the city of San Diego. The city was involved in the lawsuit because it had a contract with the contractor to install the new sewer line, and the city did periodically inspect the contractor's work. The lawsuit settled prior to trial in March 1989.

There are several important pieces of evidence that can be used to determine the cause of the cave-in. The first is the statement by the injured worker. The day after the accident, this statement was prepared:

> I [the injured worker], on 28 August, 1986 at 2:15 P.M., was down in the trench making grade. I am a pipe layer. The shores in the trench were 4 feet [1.2 m] apart. The two shores on the north of me collapsed—went

together. I was squeezed but not hurt too much. I got loose by myself and climbed out on the ladder. It happened so fast I do not know how it happened. [The employer's] equipment is in good condition. To the best of my recollection this is all I remember about this accident. The boss rushes me more than normal.

The shores described by the injured worker were made of wood with hydraulic jacks used as cross braces (struts). Apparently, utility lines ran through the trench, and thus it was not possible to use a sliding trench metal shield. The wood shoring did not extend above the ground surface because steel plates were placed over the trench in the evening to allow traffic through.

The accident was investigated by Cal/OSHA and they issued an accident investigation report ("Accident Investigation Report," no date). The report is undated but was apparently prepared on or shortly after the day of the accident. The narrative section of the report states:

> I cited the employer under section 1541 (c)(6) which requires uprights of trench shoring systems extend all the way to the bottom of the trench when running soil is encountered. I rated the violation "serious" because there was a substantial probability that it could have resulted in death or serious physical harm. Also, with reasonable diligence the employer could have known the soil around the storm drain would run.

The Cal/OSHA inspector also prepared a diagram depicting the failure of the shoring system. As shown in Fig. 6.54, the depth of the trench at the time of failure was 4.5 m (14.9 ft). The sheeting consisted of 1.9-cm (¾-in.) plywood, the uprights were 3.7 m (12 ft) long and 0.1 m by 0.35 m (4 in. by 14 in.) in cross section, and the cross braces (struts) were hydraulic jacks. As written by the Cal/OSHA inspector (Fig. 6.54), "upright slid down when sand ran from around storm drain." This loss of soil support created a void behind the north side of the shoring (directly above the storm drain).

On September 4, 1986, the contractor was issued a citation by Cal/OSHA ("Citation," 1986), with the type of alleged violation listed as serious. In the citation, Cal/OSHA stated:

> Uprights of trench shoring systems shall extend to a least the top of the trench and to as near the bottom as permitted by the material being installed, but not more than two feet [0.6 m] from the bottom. When running soil is encountered, shoring shall extend to the bottom of the trench. Twelve foot [3.7 m] long uprights were used to shore a trench that was 14.9 feet [4.5 m] deep. Running soil was encountered and the trench caved-in. The bottoms of the uprights were more than 2 feet [0.6 m] above the bottom of the trench.

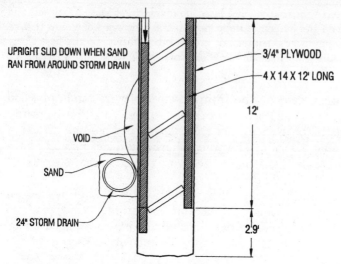

Figure 6.54 Diagram of trench cave-in (initially prepared by California/OSHA inspector, redrawn by author).

On November 24, 1986, the civil engineer who had designed the shoring system issued a report, *Slough-In Incident Report* (1986). In this report, the shoring design engineer stated, "The sloughing was apparently due to the instability of the material surrounding the parallel storm drain and the strength of the shoring system was not a factor."

In summary, there were two main lessons learned from this shoring failure:

1. *Storm drain.* The 0.6-m- (24-in.-) diameter storm drain that runs parallel to the excavated trench (see Fig. 6.54) was surrounded by sand that was described as running soil. It was this loss of soil support from the storm-drain backfill that allowed the north shoring to slide downward.

2. *Trench shoring uprights.* The trench shoring uprights (see Fig. 6.54) did not extend to the bottom of the trench. If they had, they would not have been able to slide downward and cause the collapse of the shoring system.

6.10   Dam Failures

6.10.1   Large dams

A dam failure has the potential to cause more damage and death than the failure of any other type of civil engineering structure. The worst type of failure is when the reservoir behind a large dam is full and the dam suddenly ruptures, which causes a massive flood wave to surge downstream. When this type of sudden dam failure occurs without warning, the toll can be especially high. For example, the sudden collapse in 1928 of the St. Francis Dam in California killed about 450 people, which was California's second most destructive disaster, exceeded in loss of life and property only by the San Francisco earthquake of 1906. Like many dam failures, most of the dam was washed away and the exact cause of the failure is unknown. The consensus of forensic geologists and engineers is that the failure was due to adverse geologic conditions at the site. Three distinct geologic conditions could have led to the disaster: (1) slipping of the rock beneath the easterly side of the dam along weak geologic planes; (2) slumping of rocks on the westerly side of the dam as a result of water saturation; or (3) seepage of water under pressure along a fault beneath the dam (*Committee Report*, 1928; Association of Engineering Geologists, 1978; Schlager, 1994).

On the basis of Middlebrooks' (1953) comprehensive study of earth dams, the most common causes of catastrophic failure are listed below:

Overtopping. The most frequent cause of failure is water flowing over the top of the earth dam. This generally happens during heavy or record-breaking rainfall, which causes so much water to enter the reservoir that the spillway can not handle the flow or the spillway becomes clogged. Once the earth dam is overtopped, the erosive action of the water can quickly cut through the shells and core of the dam.

Piping. The second most common cause of earth dam failure is piping (Middlebrooks, 1953). *Piping* is defined as the progressive erosion of the dam at areas of concentrated leakage. As water seeps through the earth dam, seepage forces are generated that exert a viscous drag force on the soil particles. If the forces resisting erosion are less than the seepage forces, the soil particles are washed away and the process of piping commences. The forces resisting erosion include the cohesion of the soil, interlocking of individual soil particles, confining pressure from the overlying soil, and the action of any filters (Sherard et al., 1963). Figure 6.55 shows a series of photographs of the progressive piping failure of a small earth dam (from Sherard et al., 1963). The arrows in Fig. 6.55 point to the location of the piping failure.

There can be many different reasons for the development of piping in an earth dam. Some of the more common reasons are as follows (Sherard et al., 1963):

■ Poor construction control, which can result in inadequately compacted or pervious layers in the embankment.

Figure 6.55 Progressive piping of abutment leak at small dam. 3:30 P.M.; (b) 3:45 P.M.; (c) 4:30 P.M.; (d) 5:30 P.M. [*From Sherard et al. (1963), reprinted with permission from John Wiley & Sons, Inc.*]

- Inferior compaction adjacent to concrete outlet pipes or other structures.

- Poor compaction and bond between the embankment and the foundation or abutments.

- Leakage through cracks that develop when portions of the dam are subjected to tensile strains caused by differential settlement of the dam.

- Cracking in outlet pipes, which is often caused by foundation settlement, spreading of the base of the dam, or deterioration of the pipe itself.

- Leakage through the natural foundation soils under the dam.

Leakage of the natural soils under the dam can be due to the natural variation of the foundation material. Any seepage erupting on the downstream side of the dam is likely to cause "sand boils," which are circular mounds of soil deposited as the water exits the ground surface (see Fig. 6.56). Sand boils, if unobserved or unattended, can lead to complete failure by piping (Sherard et al., 1963).

Clean (uncemented) sand is probably the most susceptible to piping, with clay being the most resistant. There can be exceptions, such as dispersive clays, which Perry (1987) defines as follows:

> Dispersive clays are a particular type of soil in which the clay fraction erodes in the presence of water by a process of deflocculation. This occurs when the interparticle forces of repulsion exceed those of attraction so that the clay particles go into suspension and, if the water is flowing such as in a crack in an earth embankment, the detached particles are carried away and piping occurs.

According to Sherard (1972), one of the largest known areas of dispersive clays in the United States is north-central Mississippi. The cause of failure for several earth dams has been attributed to the piping of dispersive clays (Bourdeaux and Imaizumi, 1977; Stapledon and Casinader, 1977; Sherard et al., 1972).

Slope instability.   Another common cause of dam failures is slope instability. There could be a gross slope failure of the upstream or downstream faces of the dam or sliding of the dam foundation. The development of these failures is similar to that of the gross slope failures and landslides discussed in Secs. 6.5 and 6.6, and slope stability analyses can be used to investigate these types of failures. These slides can be grouped into three general categories (Sherard et al., 1963):

Figure 6.56   Typical sand boil. (*From Sherard et al. (1963), reprinted with permission from John Wiley & Sons, Inc.*]

1. Slides during construction, usually involving a failure through the natural ground underlying the dam
2. Slides on the downstream slope during reservoir operation
3. Slides on the upstream slope after reservoir drawdown.

Besides the three most common causes of earth dam failure (overtopping, piping, and slope instability), there can be many other reasons for the collapse of or damage to earth dams (see Sherard et al., 1963). Because of the complexity of large dam failures, the forensic engineer will likely be a member of a committee appointed to investigate the failure. Other members of the investigating committee could include forensic geologists and hydrogeologists. The investigation of the failure may be quite extensive and could include the interviewing of witnesses, researching the design and construction documents, reviewing the data from monitoring devices, and conducting analyses of slope stability and seepage. The checklist provided in Table 6.1 may also prove useful in the forensic investigation of dam failures.

6.10.2   Small dams

Small dams have been classified as those dams having a height of less than 12 m (40 ft) or those that impound a volume of water less than 1,000,000 m³ (1000 acre-ft) (Corns, 1974). Sowers (1974) states that although failures of large dams are generally more spectacular, failures of small dams occur far more frequently. Reasons for the higher frequency of failures include: (1) lack of appropriate design, (2) owners believe that the consequences of failure of a small dam will be minimal, (3) small dam owners frequently have no previous experience with dams, and (4) smaller dams are often not maintained (Sowers, 1974). Drawing from his experience, Griffin (1974) lists some of the common deficiencies in small dams:

1. Subsurface or geological investigations were minimal or nonexistent.

2. No provision was made for future maintenance and was totally ignored for many years.

3. Slopes of many of the structures were constructed too steep for routine maintenance.

4. Construction supervision varied from inadequate to nonexistent.

5. Hydrological design was deficient in that the top elevation of the dam was set an arbitrary distance above the pool with no flood routings being made. These hydrological deficiencies are now amplified by development within the watershed.

6. Inadequate purchase of land to protect the investment in the project.

6.10.3 Landslide dams

Landslide dams usually develop when there is a blockage of a valley by landslides, debris flow, or rock fall. Concerning landslide dams, Schuster (1986) states:

> Landslide dams have proved to be both interesting natural phenomena and significant hazards in many areas of the world. A few of these blockages attain heights and volumes that rival or exceed the world's largest man-made dams. Because landslide dams are natural phenomena and thus are not subject to engineering design (although engineering methods can be utilized to alter their geometries or add physical control measures), they are vulnerable to catastrophic failure by overtopping and breaching. Some of the world's largest and most catastrophic floods have occurred because of failure of these natural dams.

According to Schuster and Costa (1986), most landslide dams are short-lived. In their study of 63 landslide dams, 22 percent failed in less than one day after formation, and half failed within 10 days. According to Schuster and Costa (1986), overtopping was by far the most frequent cause of landslide-dam failure.

Chapter

7

Other Geotechnical and Foundation Problems

The following notation is introduced in this chapter:

Symbol	Definition
H	Height of the retaining wall
k_a	Active earth pressure coefficient
P_a	Active earth pressure resultant force
P_p	Passive earth pressure resultant force
Q	Uniform surcharge pressure
W	Resultant of the vertical retaining wall loads
x'	Horizontal distance from W to the toe of the footing
Y	Horizontal displacement of the retaining wall
μ	Friction coefficient

7.1 Introduction

Chapters 4, 5, and 6 have dealt specifically with settlement (downward movement), expansive soil (upward movement), and slope movement (lateral movement). The purpose of this chapter is to discuss other geotechnical and foundation problems that are likely to be encountered by the forensic engineer. It is difficult to cover all of the situations that arise in practice and therefore the emphasis of this chapter is on the more common forensic problems. Section 7.2 presents a brief discussion of earthquakes. A structure could be damaged by the seismic energy of the earthquake or actual rupture of the

215

ground due to fault movement. Other problems caused by the ground shaking from earthquakes include the settlement of loose soil, slope movement or failure, and liquefaction of loose granular soil below the groundwater table.

Section 7.3 discusses erosion, which is caused by raindrop impact and the transportation of soil particles by flowing water. Erosion can also be human-induced, such as described in the case study which deals with the erosion of a sea cliff caused by the rupture of a storm drain line during a heavy rainfall.

Section 7.4 deals with deterioration. All man-made and natural materials are susceptible to deterioration. This topic is so broad that it is not possible to cover every type of geotechnical or foundation element susceptible to deterioration. Instead, this section will deal with three of the more common types of deterioration: sulfate attack of concrete, pavement distress, and frost related damage.

Section 7.5 presents a brief discussion of the damaging effects of tree roots. Growing tree roots develop tremendous hydraulic forces that enlarge cracks, separate bricks, and lift concrete and blacktop.

As compared to the number of structures damaged by settlement, there are much fewer structures affected by bearing capacity failures. However, a bearing capacity failure of a structure can be catastrophic, because it usually involves the sudden and complete collapse of the structure. Bearing capacity failures are discussed in Sec. 7.6.

The repair and maintenance of historic structures present some unique challenges for the forensic engineer. Historic structures are discussed in Sec. 7.7.

Section 7.8 deals with unusual soil. There are many types of unusual soils that have rare or unconventional engineering behavior. An example of unusual soil is diatomaceous earth, which contains the siliceous remains of diatoms.

The final section (7.9) is devoted to retaining walls. They have been included as a separate section because they can be damaged by different causes, such as settlement, overturning, or bearing capacity failure. Retaining wall failures caused by expansive soil backfill have been discussed in Sec. 5.3.1, and the effects of moisture on retaining or basement walls will be discussed in Sec. 8.3.

7.2 Earthquakes

Earthquakes throughout the world cause a considerable amount of death and destruction. Much as diseases will attack the weak and infirm, earthquakes damage those structures that have inherent weaknesses or age-related deterioration. Those buildings that are

Figure 7.1   Collapse of brick chimney, 1994 Northridge, California, earthquake.

nonreinforced, poorly constructed, weakened from age or rot, or
underlaid by soft or unstable soil are most susceptible to damage. For
example, Fig. 7.1 shows the collapse of a brick chimney during the
1994 Northridge, California, earthquake. The chimney was poorly
constructed, and, as shown in Fig. 7.1, there were no connections
between the chimney and house.

Besides buildings, earthquakes can damage other structures, such
as earth dams. Sherard et al. (1963) indicate that in the majority of
dams shaken by severe earthquakes, two primary types of damage
have developed: (1) longitudinal cracks at the top of the dam and (2)
crest settlement. In the case of the Sheffield Dam, a complete failure
did occur, probably due to liquefaction of the very loose and saturated
lower portion of the embankment (Sherard et al., 1963).

A structure can be damaged by many different earthquake effects.
The purpose of the following sections is to provide a brief summary of
the more common earthquake effects that could be encountered by
the forensic engineer.

7.2.1   Surface faulting and ground rupture

Surface fault rupture caused by an earthquake is important because
it severely damages buildings, bridges, dams, tunnels, canals, and

underground utilities (Lawson et al., 1908; Ambraseys, 1960; Duke, 1960; California Department of Water Resources, 1967; Bonilla, 1970; Steinbrugge, 1970). Fault displacement is defined as the relative movement of the two sides of a fault, measured in a specific direction (Bonilla, 1970). Example of large surface fault rupture are the 11 m (35 ft) of vertical displacement in the Assam earthquake of 1897 (Oldham, 1899) and the 9 m (29 ft) of horizontal movement during the Gobi-Altai earthquake of 1957 (Florensov and Solonenko, 1965). The length of the fault rupture can be quite significant. For example, the estimated length of surface faulting in the 1964 Alaska earthquake varied from 600 to 720 km (Savage and Hastie, 1966; Housner, 1970).

A recent (geologically speaking) earthquake caused the fault rupture shown in Fig. 7.2. The fault is located at the base of the Black Mountains, in California. The vertical fault displacement caused by the earthquake is the vertical distance between the two arrows in Fig. 7.2. The fault displacement occurred in an alluvial fan being deposited at the base of the Black Mountains. Most structures would be unable to accommodate the huge vertical displacement shown in Fig. 7.2.

In addition to fault rupture, there can also be ground rupture away from the main trace of the fault. These ground cracks could be caused

Figure 7.2   Fault rupture at the base of the Black Mountains (arrows indicate the amount of vertical displacement caused by the earthquake).

Figure 7.3   Ground rupture, 1994 Northridge, California, earthquake.

by many different factors, such as movement of subsidiary faults, auxiliary movement that branches off from the main fault trace, or ground rupture caused by the differential or lateral movement of underlying soil deposits. For example, Fig. 7.3 shows ground rupture during the 1994 Northridge, California, earthquake. The direction of the ground shear movement shown in Fig. 7.3 is toward the northwest. The ground movement sheared both the concrete patio and adjacent pool and knocked the house off its foundation.

When investigating damage caused by surface faulting and ground rupture, the forensic engineer will usually observe the presence of ground displacement and cracks as well as the shear type damage of structures such as shown in Figs. 7.2 and 7.3.

7.2.2 Liquefaction

The typical subsurface condition that is susceptible to liquefaction is a loose sand that has been newly deposited or placed, with a groundwater table near ground surface. During an earthquake, the ground shaking causes the loose sand to contract, resulting in an increase in pore water pressure. The increase in pore water pressure causes an upward flow of water to the ground surface, where it emerges in the form of mud spouts or sand boils. The development of high pore water pressures due to the ground shaking and the upward flow of water may turn the sand into a liquefied condition, a process which has been termed *liquefaction*. Structures on top of the loose sand deposit that has liquefied during an earthquake will sink or fall over, and buried tanks will float to the surface when the loose sand liquefies (Seed, 1970).

Liquefaction can also cause lateral movement of slopes and create flow slides (Ishihara, 1993). Seed (1970) states:

> If liquefaction occurs in or under a sloping soil mass, the entire mass will flow or translate laterally to the unsupported side in a phenomena termed a flow slide. Such slides also develop in loose, saturated, cohesionless materials during earthquakes and are reported at Chile (1960), Alaska (1964), and Niigata (1964).

Another example of lateral movement of liquefied sand is shown in Fig. 7.4 (from Kerwin and Stone, 1997). This damage occurred to a marine facility at Redondo Beach King Harbor during the 1994 Northridge, California, earthquake. The 5.5 m (18 ft) of horizontal displacement was caused by the liquefaction of an offshore sloping fill mass that was constructed as a part of the marine facility.

There can also be liquefaction of seams of loose saturated sands within a slope. This can cause the entire slope to move laterally along the liquefied layer at the base. These types of gross slope failures caused by liquefied seams of soil caused extensive damage during the 1964 Alaska earthquake (Shannon and Wilson, Inc., 1964; Hansen, 1965). It has been observed that slope movement of this type typically results in little damage to structures located on the main slide mass, but buildings located in the graben area are subjected to large differential settlements and are often completely destroyed (Seed, 1970).

Figure 7.4 Damage to marine facility, 1994 Northridge, California, earthquake. (*From Kerwin and Stone, 1997, reprinted with permission from the American Society of Civil Engineers.*)

7.2.3 Slope movement and settlement

Besides loose saturated sands, other soil conditions can result in slope movement or settlement during an earthquake. For example, Grantz et al. (1964) described an interesting case of ground vibrations from the 1964 Alaska earthquake that caused 0.8 m (2.5 ft) of alluvium settlement. Other loose soils, such as cohesionless sand and gravel, will also be susceptible to settlement due to the ground vibrations from earthquakes.

Slopes having a low factor of safety can experience large horizontal movement during an earthquake. Types of slopes most susceptible to movement during earthquakes include those slopes composed of soil that loses shear strength with strain (such as sensitive soil) or ancient landslides that can become reactivated by seismic forces (Day and Poland, 1996).

7.2.4 Translation and rotation

An unusual feature of earthquakes is translation and rotation of objects. For example, Fig. 7.5 shows a photograph of a brick mailbox that rotated and translated (moved laterally) during the Northridge

Figure 7.5   Rotation of brick mailbox, 1994 Northridge, California, earthquake.

earthquake. The "initial position" in Fig. 7.5 refers to the pre-earthquake position of the mailbox.

Earthquakes have caused the rotation of other movable objects, such as grave markers (Athanasopoulos, 1995; Yegian et al., 1994). According to Athanasopoulos (1995), such objects will rotate in such a manner as to be aligned with the strong component of the earthquake. Besides rotation, translation (lateral movement) can also occur during an earthquake. The objects will tend to move in the same direction as the propagation of energy waves, i.e., in a direction away from the epicenter of the earthquake.

7.2.5   Foundation behavior

The type of foundation can be very important in the performance of the structure during the earthquake. Foundations such as mats or posttensioned slabs may enable the building to remain intact, even with substantial movements. This section presents a discussion of the effect of foundation type on the damage to single-family houses during the Northridge, California, earthquake. The emphasis of this section is on earthquake-related foundation damage and not foundation damage caused by earthquake-induced soil movement.

The Northridge earthquake occurred on January 17, 1994 at 4:31 A.M. The earthquake magnitude was 6.7. There was collapse of specially designed structures such as multistory buildings, parking garages, and freeways. In some areas, the most severe damage would indicate a Modified Mercalli (MM) intensity of IX, although MM VII to VIII was more widespread. Because the Northridge earthquake occurred in a suburban community, damage to single-family houses was common.

Damage estimates varied considerably. For both public and private facilities, the total cost of the Northridge earthquake was on the order of $20 to $25 billion. This makes the Northridge earthquake California's most expensive natural disaster. Given the significant damage caused by this earthquake, the number of deaths was relatively low. This was partly because most people were asleep at home at the time of the earthquake.

The Northridge earthquake data was obtained from site investigations performed by the author in 1994. This work was requested by several different insurance companies. Most of the residences investigated were single-family detached houses having wood-frame construction with exterior stucco and interior wallboard or lath and plaster.

For those locations with similar earthquake intensity and duration, the type of foundation was a factor in the severity of damage. For most single-family houses in southern California, the foundations are either raised wood floor or slab-on-grade. This is because frost heave is not a concern, because of the climate, and the foundations tend to be of shallow depth.

Raised wood floor with isolated interior posts. This type of raised wood floor foundation consists of continuous concrete perimeter footings and interior (isolated) concrete pads. The floor beams span between the continuous perimeter footings and the isolated interior pads. The continuous concrete perimeter footings are typically constructed so that they protrude about 0.3 to 0.6 m (1 to 2 ft) above adjacent pad grade. The interior concrete pad footings are not as high as the perimeter footings, and short wood posts are used to support the floor beams. The perimeter footings and interior posts elevate the wood floor and provide for a crawl space below the floor.

In southern California, the raised wood floor foundation having isolated interior pads is common for houses 30 years or older. Most newer houses are not constructed with this foundation type. In some cases, the raised wood floor was not adequately bolted to the concrete foundation. Only a few bolts, or in some cases nails, were used to attach the wood sill plate to the concrete foundation.

In general, damage was more severe to houses having this type of raised wood floor foundation. There may be several different reasons for this behavior:

1. *Lack of shear resistance of wood posts.* As previously mentioned, in the interior, the raised wood floor beams are supported by short wood posts bearing on interior concrete pads. During the earthquake, these short posts were vulnerable to collapse or tilting.

2. *No bolts or inadequate bolted condition.* Because in many cases the house was not adequately bolted to the foundation, it slid or even fell off the foundation during the earthquake. In other cases the bolts were spaced too far apart and the wood sill plate split, allowing the house to slide off the foundation.

3. *Age of residence.* The houses having this type of raised wood floor foundation were older. The wood was more brittle and in some cases weakened due to rot or termite damage. In some cases, the concrete perimeter footings were nonreinforced or had weakened due to prior soil movement, making them more susceptible to cracking during the earthquake.

Slab-on-grade. In southern California, the concrete slab-on-grade is the most common type of foundation for houses constructed within the past 20 years. It consists of perimeter and interior continuous footings, interconnected by a slab-on-grade. Construction of the slab-on-grade begins with the excavation of the interior and perimeter continuous footings. Steel reinforcing bars are commonly centered in the footing excavations and a wire mesh is used as reinforcement for the slab. The concrete for both the footings and the slab is usually placed at the same time, in order to create a monolithic foundation. Unlike the raised wood floor foundation, the slab-on-grade does not have a crawl space.

In general, for those houses with a slab-on-grade, the wood sill plate was observed to be bolted to the concrete foundation. In many cases, the earthquake caused the development of an exterior crack in the stucco at the location where the sill plate meets the concrete foundation. In some cases, the crack could be found on all four sides of the house. The crack developed when the house framing bent back and forth during the seismic shaking.

Of the two foundations subjected to similar earthquake intensity and duration, those houses having a slab-on-grade generally had the best performance. This is because the slab-on-grade is typically stronger due to steel reinforcement and monolithic construction, the houses are newer (less wood rot and concrete deterioration), there is greater frame resistance because of the construction of shear walls, and the wood sill plate is in continuous contact with the concrete foundation.

It should be mentioned that although the slab-on-grade generally had the best performance, there were such houses that were nevertheless severely damaged. In many cases, these houses did not have adequate shear walls, there were numerous wall openings, or there was poor construction. The construction of a slab-on-grade by itself was not enough to protect the structure from collapse if the structural frame above the slab did not have adequate shear resistance.

Besides investigating earthquake-related damage, the forensic engineer could also be involved with the retrofitting of existing structures. For the two foundation types previously discussed, the raised wood floor with isolated posts is rarely used for new construction. But there are numerous older houses that have this foundation type and in many cases, the wood sill plate is inadequately bolted to the foundation. Bolts or tie-down anchors could be installed to securely attach the wood framing to the concrete foundation. Wood bracing or plywood could be added to the open areas between posts to give the foundation more shear resistance.

In the Northridge earthquake, the observed foundation damage indicated the importance of tying together the various foundation elements. To resist damage during the earthquake, the foundation should be monolithic with no gaps in the footings or planes of weakness due to free-floating slabs.

7.3 Erosion

The process of slope erosion can begin as slopewash. The term *slopewash* refers to one form of erosion in which water moves as a thin and relatively uniform film (Rice, 1988). With time, the flow of water may concentrate into a series of slightly deeper pathways (called *rills*) and this is known as *rillwash*. With continued erosion, gullies could emerge and ultimately there could be the formation of channeled stream flow through the slope. Table 7.1 lists five levels of slope erosion and Fig. 7.6 shows severe erosion of a slope.

The factors that affect sheet and rill erosion have been discussed by Smith and Wischmeier (1957). They indicate that there are two principal processes of sheet erosion: raindrop impact and transportation of soil particles by flowing water. According to Smith and Wischmeier (1957), the amount of soil loss is governed by six factors: length of slope, slope gradient, ground cover, soil type, management, and rainfall. Soil type would include such factors as type and degree of cementation of particles.

TABLE 7.1   Level of Erosion

Level of erosion (1)	Classification (2)	Description (3)
1	Very slight	Minor erosion; at the base of the slope, minor accumulation of debris.
2	Slight	Erosion consists of rills, which may be up to about 8 cm deep; some debris at the base of the slope.
3	Appreciable	Rills up to about 0.3 m deep. Debris at the base of the slope.
4	Severe	Rills from about 0.3 to 1 m deep and gullies are beginning to form; considerable debris at the base of the slope.
5	Very severe	Deep erosion channels, consisting of rills and gullies; development of pipes causing underground erosion; very large accumulation of debris at the base of the slope.

Figure 7.6 Example of severe erosion of a slope (note broken drainage ditch in lower left corner of photograph).

As previously mentioned in Sec. 6.10, clean (uncemented) sand is most susceptible to erosion and clay is the most resistant, with certain exceptions, such as dispersive clays. McElroy (1987) describes unique erosion features caused by dispersive clays. For example, the erosion of fill or cut slopes can take the form of vertical or near-vertical tunnels called "jugs." McElroy (1987) states that these jugs may develop as a result of small drying cracks, rodent holes, openings created by decaying roots, animal and vehicle tracks, or other small surface depressions that permit rainfall or surface water to collect. The jugs frequently consist of a small hole (often less than 25 mm) on the surface of the cut or fill slope, and this hole can extend to a depth of 3 m (10 ft) with a bottom diameter of 1 m (3 ft). When these jugs collapse, the slope can erode to form severe rills and gullies.

7.3.1 Sea cliffs

Sea cliffs can be more susceptible to erosion because of a lack of vegetation caused by salt accumulation in the soil from sea spray. A reduction in vegetation will provide less rain impact protection and less root reinforcement to hold the soil in place. Another factor that causes sea cliff erosion that is unique to the ocean environment is toe erosion due to ocean wave impact and scour.

Figure 7.7 shows a photograph of various sea cliff erosion control measures, in Solana Beach, California. The sea cliff shown in Fig. 7.7 consists of weakly cemented to moderately cemented sandstone. The left side of the photograph shows the natural slope with no irrigation, except perhaps at the top of slope. Notice that the erosion of the face of the slope has created numerous rills. The sea cliff protection consists predominantly of boulder riprap placed at the toe of the slope.

In the middle of Fig. 7.7, a second (more rigorous) system of sea cliff protection has been constructed. At the toe of the slope, there is a concrete sea wall to protect the slope from ocean waves. Above the sea wall, a crib wall has been constructed which supports a cement-treated slope. The entire sea cliff face has been essentially rebuilt with a more resistant, human-made erosion control system.

At the right side of Fig. 7.7, there is the sea wall to again protect the toe of the slope from ocean waves. But above the sea wall, there is extensive planting and irrigation to promote the growth of vegetation. The slope face is essentially completely vegetated.

Figure 7.8 shows other commonly used erosion control measures for sea cliffs in Solana Beach, California. The left side of Fig. 7.8 shows a gunite retaining structure. Typically the gunite is anchored to the

Figure 7.7   Sea cliff erosion protection.

Figure 7.8   Sea cliff erosion protection.

Figure 7.9 Sea cliff erosion protection.

rock through the use of tieback anchors. On the right side of Fig. 7.8, segmented multilevel retaining structures, with reinforced earth backfill, have been constructed to protect the toe and slope face from erosion.

As a localized stabilization measure, the caves or gullies can be filled with grout or gunite. Figure 7.9 shows an example of this erosion control measure. Note in Fig. 7.9 that a lowering of the beach has exposed the bottom of the grout. Undermining of the grout is one disadvantage of this erosion control measure.

Case study. The purpose of this section is to describe a case study of sea cliff erosion due to the failure of a storm drain system in Encinitas, California. A site plan sketch of the sea cliff is shown in Fig. 7.10. As shown in this figure, a 0.46-m- (18-in.-) diameter corrugated metal storm drain was used to drain a portion of Neptune Avenue. Water entered the storm drain at a collection box on the west side of Neptune Avenue. Once water entered the collection box and the corrugated storm drain line, it traveled across the building lot, down the sea cliff, and then discharged from the storm drain line onto the beach. The storm drain line was buried, except at the toe of the sea cliff.

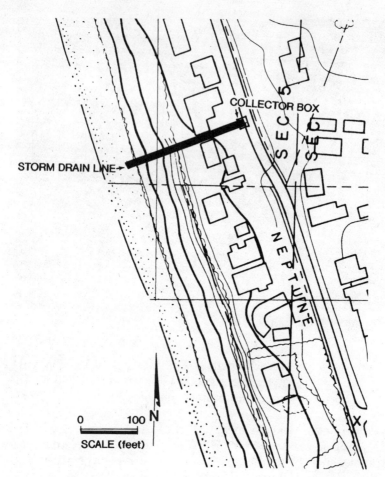

STORM DRAIN LINE→

COLLECTOR BOX

0   100 N

SCALE (feet)

Figure 7.10   Site plan.

The sea cliff at the location of the storm drain is composed of Pleistocene and Eocene Age terrace deposits. In particular, the sea cliff is composed of weakly to moderately cemented sandstone. The overall stability of the sea cliff is generally stable with no out-of-slope bedding planes. The height of the sea cliff is about 23 m (75 ft) with a predominant slope inclination of about 45°.

At two different locations, specimens were obtained by manually driving Shelby tubes into the sandstone. The sandstone was extruded from the Shelby tubes and the specimens were subjected to an index test for erosion potential (Day, 1990b). The results indicate that the near surface sandstone has a "very severe" potential for erosion.

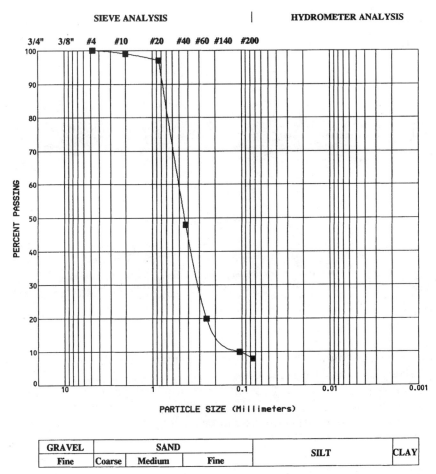

Figure 7.11   Grain size distribution.

Figure 7.11 shows the grain size distribution curve for the mineral grains composing the sandstone. The poorly graded sandstone predominantly contains fine to medium-size sand grains with about 8 percent silt size particles. The dry density and moisture content of the sandstone are 1.7 Mg/m^3 (104 pcf) and 2 percent, respectively.

The 0.46-m- (18-in.-) diameter corrugated storm drain line failed during a rainstorm on March 20–21, 1991. The recorded rainfall was 74 mm (2.9 in.). The storm drain broke at a location about halfway down the sea cliff. The cause of the pipe failure was deterioration (corrosion) which had weakened the pipe. The exact age of the storm

Figure 7.12   Location of pipe break.

drain was unknown, but the adjacent homeowners indicated it was probably 50 to 60 years old.

Figure 7.12 shows the location of the pipe break. As a temporary measure, after the pipe broke, a plastic pipe was inserted over the end of the corrugated metal pipe in order to reduce the erosion of the sea cliff.

Figure 7.13 shows two views, one from the top of slope and the other from the beach, of the sea cliff erosion caused by the pipe break. As shown in Figs. 7.13a and 7.13b, when the pipe broke, water removed the vegetation that acted as a protective cover. Then the water easily eroded into the sea cliff because of its "very severe" erosion potential. Most of the eroded material either washed into the Pacific Ocean or was deposited at the beach where it was eventually removed by beach erosion. The estimated volume of the eroded material (i.e., the size of the depression in the sea cliff) is 300 m³ (10,000 ft³).

In summary, the weakly cemented sandstone that composed the sea cliff tested as having a "very severe" erosion potential. Once the protective vegetation cover was eliminated, water from the pipe leak easily eroded the sea cliff as shown in Figs. 7.13a and 7.13b. The failure of the storm drain could probably have been prevented if it had been inspected and periodically maintained.

Figure 7.13a Erosion due to pipe leak.

Figure 7.13b Erosion due to pipe leak.

7.3.2   Badlands

Badlands have been defined as follows (Stokes and Varnes, 1955):

> An area, large or small, characterized by extremely intricate and sharp erosional sculpture. Badlands usually develop in areas of soft sedimentary rocks such as shale, but may also occur in decomposed igneous rocks, loess, etc. The divides are sharp and the slopes are scored by intricate systems of ravines and furrows. Fantastic erosional forms are commonly developed through the unequal erosion of hard and soft layers. Vegetation is scanty or lacking and there is a notable lack of coarse detritus. Badlands occur chiefly in arid or semiarid climates where the rainfall is concentrated in sudden heavy showers. They may, however, occur in humid regions where vegetation has been destroyed, or where soil and coarse detritus are lacking.

Figure 7.14 presents two views of badlands near Death Valley, California. The arrows in Figs. 7.14a and 7.14b point to the remains of an asphalt roadway located in Golden Canyon. According to the National Park Service, a 4-day rainstorm in February 1976 dropped 5.8 cm (2.3 in.) of rain in this area. There was tremendous runoff that undermined and eroded away the pavement as shown in Figs. 7.14a and 7.14b. Such intense erosive action can easily wash away any vegetation that may have gained a foothold in the badlands.

Figure 7.15 shows erosion of a debris flow. The cap rock has protected the underlying debris flow material from erosion and created this unusual landform.

7.4   Deterioration

In regard to deterioration, the National Science Foundation (NSF, 1992) states:

> The infrastructure deteriorates with time, due to aging of the materials, excessive use, overloading, climatic conditions, lack of sufficient maintenance, and difficulties encountered in proper inspection methods. All of these factors contribute to the obsolescence of the structural system as a whole. As a result, repair, retrofit, rehabilitation, and replacement become necessary actions to be taken to insure the safety of the public.

The purpose of this section is to discuss three of the more common types of deterioration: sulfate attack of concrete, pavement distress, and frost-related damage.

7.4.1   Sulfate attack

Sulfate attack of concrete is defined as a chemical and/or physical reaction between sulfates (usually in the soil or groundwater) and

Figure 7.14a Badlands (arrow points to remains of the asphalt pavement).

concrete or mortar, primarily with hydrated calcium aluminate in the cement-paste matrix, often causing deterioration (ACI, 1990). Sulfate attack of concrete occurs throughout the world, especially in arid areas, such as the southwestern United States. In arid regions, the salts are drawn up into the concrete and then deposited on the concrete surface as the groundwater evaporates, as shown in Figs. 7.16a and 7.16b.

Typically the geotechnical engineer obtains the representative soil or groundwater samples to be tested for sulfate content. The geotechnical engineer can analyze the soil samples or groundwater in house, or send the samples to a chemical laboratory for testing. There are different methods to determine the soluble sulfate content in soil or

Figure 7.14b Badlands (arrow points to remains of the asphalt pavement).

groundwater. One method is to precipitate out and then weigh the sulfate compounds. A faster and easier method is to add barium chloride to the solution and then compare the turbidity (relative cloudiness) of barium sulfate with known concentration standards.

Once the soluble sulfate content has been determined, the geotechnical or foundation engineer can then recommend measures, such as using type V cement, to mitigate the effects of sulfate on the concrete. If there is deterioration of a concrete foundation or retaining wall, the forensic geotechnical engineer and a concrete material specialist will frequently be part of the investigative team.

There has been considerable research, testing, and chemical analysis of sulfate attack. Two different mechanisms of sulfate attack have been discovered: chemical reactions and the physical growth of crystals.

Figure 7.15 Badlands (a cap rock has prevented erosion of the underlying debris flow).

Chemical reactions. The chemical reactions involving sulfate attack of concrete are complex. Studies (Lea, 1971; Mehta, 1976) have discovered two main chemical reactions. The first is a chemical reaction of sulfate and calcium hydroxide (which was generated during the hydration of the cement) to form calcium sulfate, commonly known as *gypsum*. The second is a chemical reaction of gypsum and hydrated calcium aluminate to form calcium sulfoaluminate, commonly called *ettringite* (ACI, 1990). As with many chemical reactions, the final product of ettringite causes an increase in volume of the concrete. Hurst (1968) indicates that the chemical reactions produce a compound of twice the volume of the original tricalcium aluminate compound. Concrete has a low tensile strength and thus the increase in

Figure 7.16a Concrete sidewalk deterioration, Mojave Desert (salts were deposited on concrete surface from evaporating groundwater).

volume fractures the concrete, allowing more sulfates to penetrate the concrete, resulting in accelerated deterioration.

Physical growth of crystals. The physical reaction of sulfate has been studied by Tuthill (1966) and Reading (1975). They conclude that there can be crystallization of the sulfate salts in the pores of the concrete. The growth of crystals exerts expansive forces within the concrete, causing flaking and spalling of the outer concrete surface. Besides sulfate, the concrete, if porous enough, can be disintegrated by the expansive force exerted by the crystallization of almost any salt in its pores (Tuthill, 1966; Reading, 1975). Damage due to crystallization of salt is commonly observed in areas where water is

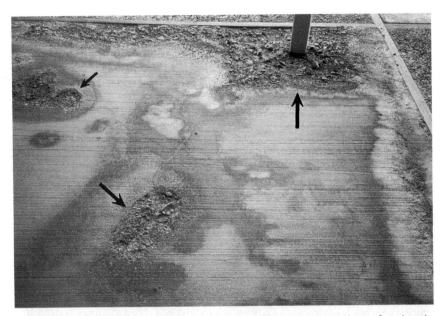

Figure 7.16b Concrete patio deterioration, Mojave Desert (arrows point to deterioration of concrete surface).

migrating through the concrete and then evaporating at the concrete surface. Examples include the surfaces of concrete dams, basement and retaining walls that lack proper waterproofing, and concrete structures that are partially immersed in salt-bearing water (such as seawater) or soils.

The forensic engineer can recognize sulfate attack by the physical loss of concrete (Fig. 7.17) or the unusual cracking and discoloration of concrete such as shown in Fig. 7.18. When investigating concrete deterioration, the forensic engineer should be aware of the factors that cause sulfate attack. In general, the degree of sulfate attack of concrete will depend on the type of cement used, quality of the concrete, soluble sulfate concentration that is in contact with the concrete, and the surface preparation of the concrete (Mather, 1968).

1. *Type of cement.* There is a correlation between the sulfate resistance of cement and its tricalcium aluminate content. As previously discussed, it is the chemical reaction of hydrated calcium aluminate and gypsum that forms ettringite. Therefore, limiting the tricalcium aluminate content of cement reduces the potential for the formation of ettringite. It has been stated that the tricalcium aluminate content of the cement is the greatest single factor that influences the resistance of concrete to sulfate attack, where in general, the lower the tri-

Figure 7.17 Physical loss of concrete due to sulfate attack.

calcium aluminate content, the greater the sulfate resistance (Bellport, 1968). Of the types of Portland cements, the most resistant cement is type V, in which the tricalcium aluminate content must be less than 5 percent. Both the ACI (ACI, 1990) and the Portland Cement Association (*Design,* 1988) have identical requirements for normal-weight concrete subjected to sulfate attack. Depending on the percentage of soluble sulfate in the soil or groundwater, a certain cement type is required. In an investigation of damage due to sulfate deterioration, the forensic engineer should compare the requirements of ACI with the actual cement type used for the concrete.

2. *Quality of concrete.* The condition of the concrete in terms of its permeability should be evaluated by the forensic engineer. In general, the more impermeable the concrete, the more difficult for the water-borne sulfate to penetrate the concrete surface. To have a low permeability, the concrete must be dense, have a high cement content, and a low water-cement ratio. Using a low water-cement ratio decreases the permeability of mature concrete (*Design,* 1988). A low water-cement

Figure 7.18 Cracking and discoloration of concrete due to sulfate attack.

ratio is a requirement of ACI (1990) for concrete subjected to soluble sulfate in the soil or groundwater. For example, the water-cement ratio must be equal to or less than 0.45 for concrete exposed to severe or very severe sulfate exposure. There are many other conditions that can affect the quality of the concrete. For example, a lack of proper consolidation of the concrete can result in excessive voids. Another condition is the corrosion of reinforcement, which may crack the concrete and increase its permeability. Cracking of concrete may also occur when structural members are subjected to bending stresses. For example, the tensile stress due to a bending moment in a footing may cause the development of microcracks, which increase the permeability of the concrete.

3. *Soluble sulfate concentration.*   When investigating sulfate attack of concrete, the forensic engineer should determine if the soil or water in contact with the concrete was tested to determine the soluble sulfate content. In some cases, the soluble sulfate may become concentrated on crack faces. For example, water evaporating through cracks in concrete flatwork will deposit the sulfate on the crack faces. This concentration of sulfate may cause accelerated deterioration of the concrete.

4. *Surface preparation of concrete.*   An important factor in concrete resistance is the surface preparation, such as the amount of cur-

ing of the concrete. Curing results in a stronger and more impermeable concrete (*Design,* 1988), which is better able to resist the effects
of salt intrusion.

Case study.   This case study deals with sulfate attack of concrete flatwork at a 150-unit condominium project located in San Diego,
California. The condominiums were constructed in 1977–78. Around
1985, the homeowners association initiated a lawsuit claiming there
were construction defects at the site. The author was retained as an
expert for the homeowners association. The homeowners association
accepted an out-of-court settlement in early 1988. It is not known
what repairs, if any, have been performed by the homeowners association after the lawsuit settled.

One area of investigation by the author involved cracking of the
concrete flatwork, especially the driveways. Figure 7.19 shows a photograph of typical cracking to the concrete driveways. Cracks of variable width are common in concrete. However, as shown in Fig. 7.19
(small arrows), the cracking is unusual in that there are wide gaps
along the cracks.

Figure 7.19 also shows the location of a 150-mm- (6-in.-) diameter
core (location of large arrow). The concrete was about 90 mm (3.5 in.)

Figure 7.19   Driveway deterioration.

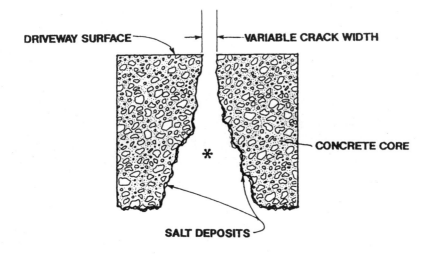

DRIVEWAY SURFACE

VARIABLE CRACK WIDTH

CONCRETE CORE

SALT DEPOSITS

✻ AREA OF DETERIORATION

Figure 7.20   Diagram of driveway deterioration.

thick, nonreinforced, and constructed of type I cement. This and other concrete cores of the driveways revealed that portions of the concrete paste had been disintegrated, leaving pieces of the coarse aggregate. Figure 7.20 presents an illustration of the typical conditions of the concrete cores. There was a triangular piece of concrete, located directly beneath the surface crack, that had disintegrated. On these surfaces, there was a heavy salt deposit. It appeared that the gaps (Fig. 7.19) were due to chemical and physical attack that had widened the surface concrete cracks. The bottom of the concrete driveway, at areas away from the cracks, appeared to be relatively intact and undamaged.

The concrete cores also revealed that the driveway had been constructed directly on the fill (there was no base below the driveways). The fill was classified as a silty clay that tested as having a highly variable soluble sulfate content. This is typical of many soil deposits containing soluble sulfate. Six soil specimens were selected for sulfate analyses. Two of the soil specimens tested as having negligible soluble sulfate, but the other four soil specimens had between 0.29 and 1.26 percent soluble sulfate, indicating "severe" sulfate exposure. The source rock that the fill was derived from did contain lenses of naturally precipitated gypsum, and the variable sulfate content was probably due to fill having different concentrations of this gypsum.

Based on the results of the forensic investigation, the deterioration mechanism of the concrete driveways was as follows:

1. Moisture migrated to underneath the concrete driveways. Because of the silty clay fill, moisture would tend to accumulate underneath the driveway slabs by capillary action in the soil. Moisture also migrates to cooler areas, such as beneath concrete slabs, due to thermo-osmosis.

2. The moisture beneath the concrete driveways caused expansion of the silty clay fill. The silty clay was generally classified as being medium to highly expansive. At some locations, differential heave at concrete joints was observed. This expansion of the clayey fill would have cracked the nonreinforced concrete. Some of the moisture that accumulated beneath the driveway also evaporated up through the concrete driveway cracks.

3. As the water evaporated at the concrete driveway cracks, the salts were precipitated on the concrete crack faces. The salts caused both physical and chemical deterioration of the concrete. This resulted in a widening of the cracks which then allowed for more moisture to evaporate though the concrete cracks.

4. Those concrete driveway cracks that developed first or had the highest concentration of sulfates developed the widest surface cracks (small arrows in Fig. 7.19).

7.4.2 Pavement deterioration

There can be many types of pavement deterioration or failure, including rutting, alligator cracking, bleeding, block cracking, raveling, corrugation, and the development of potholes or depressions. Descriptions and photographs of these types of pavement deterioration are presented by ASTM (e.g., ASTM D 5340-93 and ASTM E 1778-96a, ASTM 1997b and 1997c).

The damage and deterioration of pavements caused by expansion of a clay subgrade has been discussed in Sec. 5.5. The detrimental effects of water trapped in slow draining pavement and base will be discussed in Sec. 8.1, "Groundwater." Besides expansive soil and groundwater, there can be other factors that contribute to pavement deterioration. Probably the most common causes of premature pavement deterioration or failure are from heavier-than-expected traffic loads or an unanticipated higher volume of traffic. As mentioned in Sec. 5.5, one method for the design of flexible pavements is to use the traffic index (TI). When investigating pavement deterioration or failure, the forensic engineer should compare the actual traffic loads with the design assumptions. Besides increased traffic loads or volume, there could be many factors that contribute to the premature deterioration of pavements. Some examples include a weaker pavement surface, base course, or subgrade than assumed during the design phase. For example, the forensic engineer may discover that the actual California bearing ratio (CBR) or the R-value of the aggregate base is much less than the value assumed during the design phase, because of inadequate compaction, particle degradation, or the presence of groundwater trapped in the base course. The forensic engineer should also compare the thickness of the pavement surface and base with the original design specifications. A common cause of pavement deterioration or failure is a pavement section thickness that is much thinner than the original recommendations. Other factors to be considered in the investigation of pavement deterioration include (NAVFAC DM-21.3, 1978):

■ Characteristics, strength, and in-place density of the subgrade, base, and asphalt or concrete surface.

■ Seasonal fluctuations of groundwater and effectiveness of pavement drainage.

■ Frost susceptibility of the pavement section and the effect of freeze-thaw conditions on the subgrade.

■ Presence of weak or compressible layers in the subgrade.

■ Variability of the subgrade, which may cause differential surface movements.

7.4.3 Frost

There have been extensive studies on the detrimental effects of frost (Casagrande, 1932; Kaplar, 1970; Yong and Warkentin, 1975; and Reed et al., 1979). Two common types of damage related to frost are (1) freezing of water in cracks and (2) formation of ice lenses. In many cases, deterioration or damage is not evident until the frost has melted. In these instances, it may be difficult for the forensic engineer to conclude that frost was the primary cause of the deterioration.

Freezing of water in cracks. There is about a 10 percent increase in volume of water when it freezes, and this volumetric expansion of water upon freezing can cause deterioration or damage to many different types of materials. Common examples include rock slopes and concrete, as discussed below.

- *Rock slopes.* The expansive force of freezing water results in a deterioration of the rock mass, additional fractures, and added driving (destabilizing) forces. Feld and Carper (1997) describe several rock slope failures caused by freezing water, such as the February 1957 failure where 900 Mg (1000 tons) of rock fell out of the slope along the New York State Thruway, closing all three southbound lanes north of Yonkers.

- *Concrete.* Durability is defined by the American Concrete Institute as the ability to resist weathering, chemical attack, abrasion, or any other type of deterioration (ACI, 1982). Durability is affected by strength, but also by density, permeability, air entrainment, dimensional stability, characteristics and proportions of constituent materials, and construction quality (Feld and Carper, 1997). Durability is harmed by freezing and thawing, sulfate attack, corrosion of reinforcing steel, and reactions between the various constituents of the cements and aggregates. Damage to concrete caused by freezing could occur during the original placement of the concrete or after it has hardened. To prevent damage during placement, it is important that the fresh concrete not be allowed to freeze. Air-entraining admixtures can be added to the concrete mixture to help protect the hardened concrete from freeze-thaw deterioration.

Formation of ice lenses. Frost penetration and the formation of ice lenses in the soil frequently damage shallow foundations and pavements. The frost penetration will cause heave of the structure if moisture is available to form ice lenses in the underlying soil. The spring thaw will then melt the ice, resulting in settlement of the foundation or a weakened subgrade that will make the pavement surface susceptible to deterioration or failure. Damage to highways in the United States and Canada because of frost action is estimated to amount to millions of dollars annually (Holtz and Kovacs, 1981).

It is well known that silty soils are more likely to form ice lenses because of their high capillarity and sufficient permeability that enables them to draw up moisture to the ice lenses. When dealing with buildings possibly damaged by frost action, the forensic engineer should determine if the outside columns or walls of the building are located below the level to which frost could have caused perceptible heave.

Feld and Carper (1997) describe several interesting cases of damage due to frost action. At Fredonia, New York, the frost from a deep-freeze storage facility froze the soil and heaved the foundations upward 100 mm (4 in.). A system of electrical wire heating was installed to maintain soil volume stability.

Another case involved an extremely cold winter in Chicago, where frost penetrated below an underground garage and broke a buried sprinkler line. This caused an ice buildup which heaved the structure above the street level and sheared off several supporting columns.

7.5 Tree Roots

Damage to structures caused by tree roots is very common. Damage usually occurs to lightly loaded structures such as sidewalks, patios, roads, and block walls, where the physical increase in size of growing roots causes uplift and differential movement. There has been a considerable amount written on the destructive effects of tree roots. For example, it has been stated (Perry and Merschel, 1987):

> The most destructive weapon in a plant's arsenal is its roots. These find their way into microscopic cracks where the long process of leveling cities begins. Roots, even when they are small, develop tremendous hydraulic forces that enlarge cracks, separate bricks and lift concrete and blacktop.

Figure 7.21 presents a photograph of the most common type of damage due to tree roots. In many areas of southern California, land is expensive and most of the building pad is occupied by houses and associated roads and flatwork. The space allocated to plants can be small, such as designated planter areas around the perimeter of the house. As the trees mature, the root systems expand underneath the flatwork. As shown in Fig. 7.21, the tree root has grown beneath the concrete sidewalk, causing cracking and differential movement of the concrete sidewalk.

Figure 7.22 shows damage to a road caused by growing tree roots. Subsurface exploration revealed that the road was constructed by placing asphalt concrete atop the outer edge of the paved shoulder. This allowed the base to be in direct contact with the soil from the planter areas. The curbs in this case did not act as a barrier to restrict tree roots from gaining direct access to the roadway base. As shown in Fig. 7.22, the roots penetrated the base, and as they grew, the asphalt curb and the roadway were uplifted and cracked. Note in Fig. 7.22 that the damage was done by a eucalyptus tree of a relatively young age.

A third example of uplift due to growing tree roots is shown in Fig. 7.23. A eucalyptus tree, which is visible along the right side of the photograph, was planted in close proximity to a condominium. At this particular condominium, subsurface exploration revealed that about

Figure 7.21 Sidewalk uplift due to tree root.

0.6 m (2 ft) of fill was placed over hard, dense sedimentary rock. The roots could not easily penetrate the sedimentary rock, but rather grew in the upper zone of fill. Pipe leaks underneath the condominium slab provided a source of water and the tree roots penetrated beneath the foundation. The thin veneer of fill over hard sedimentary bedrock did not provide a zone of deformable material, and as the tree roots grew, the condominium was uplifted.

For the three examples of tree-root uplift, it was observed that damage more frequently occurs to lightly loaded structures, such as

Figure 7.22a Condition prior to test pit excavation.

sidewalks, roads, and block walls. Damage seems to be less severe for soft or loose soils that deform as the roots grow, rather than the dense or hard soils such as the base in Fig. 7.22.

7.6   Bearing Capacity Failures

7.6.1   Buildings

A general bearing capacity failure of a foundation involves shear displacement and rupture of the soil underlying the structure. Compared to the number of structures damaged by settlement, there are far fewer structures that are impacted by bearing capacity failures. This is probably because there have been extensive studies of such failures which have led to the development of bearing capacity

Figure 7.22*b* Condition after test pit excavation.

equations (e.g., Sec. 14.3 in Lambe and Whitman, 1969) that are routinely used to determine the ultimate bearing capacity of the foundation. In addition, many building codes have minimum footing dimensions and maximum allowable bearing pressures for different soil and rock conditions, such as Tables 18-I-A and 18-I-C of the *Uniform Building Code* (1997), that generally have adequate factors of safety to prevent bearing capacity failures.

When investigating bearing capacity failures of buildings, the forensic engineer will need to perform subsurface exploration and laboratory testing to determine the shear strength of the soil underlying the structure. This shear strength should be compared to the values used for the foundation design. Another important parameter to be determined is the structural load at the time of failure. This load should also be compared with the values assumed during the design

Figure 7.23   Condominium uplift.

phase. Common causes of bearing capacity failures are that the shear strength of the underlying soil was overestimated or the actual structural loads at the time of failure were greater than those assumed during the design phase.

7.6.2 Roads

Besides buildings, other structures can be susceptible to bearing capacity failures. For example, unpaved roads and roads with an inadequate pavement section or weak subgrade can also be susceptible to bearing capacity failures caused by heavy wheel loads. The heavy wheel loads can cause a general bearing capacity failure or a punching-type shear failure. These bearing capacity failures are commonly known as *rutting,* and they develop when the unpaved road or weak pavement section is unable to support the heavy wheel load.

7.6.3 Pumping of clay

Another form of bearing capacity failure is the pumping of wet clay during compaction. A commonly used definition of *pumping* is the softening and squeezing of clay from underneath the compaction equipment. Continual passes of the compaction equipment can cause a decrease in the undrained shear strength of the wet clay and the pumping may progressively worsen. Figures 7.24 and 7.25 show the

Figure 7.24 Bearing capacity failure during compaction (pumping clay).

pumping of a wet clay subgrade. Pumping is dependent on the penetration resistance of the compacted clay. Turnbull and Foster (1956) present data on the California bearing ratio of compacted clay, and they show that the penetration resistance approaches zero when the clay is compacted wet of optimum (i.e., the clay can exhibit pumping).

There are many different methods to stabilize pumping soil. Commonly used methods are field drying of the clay, adding a chemical agent (such as lime) to the clay, or placing a geotextile on top of the pumping clay to stabilize its surface (Winterkorn and Fang, 1975). Another common procedure to stabilize pumping clay is to add gravel to the clay. The typical procedure is to dump angular gravel at ground surface and then work it in from the surface. The angular gravel produces a granular skeleton, which then increases both the undrained shear strength and penetration resistance of the mixture.

7.7 Historic Structures

The repair and maintenance of historic structures present unique challenges to the geotechnical and foundation engineer. The forensic engineer could be involved in many different types of problems with historic structures. Common problems include structural weakening and deterioration due to age or environmental conditions, original poor construction practices, inadequate design, and faulty mainte-

Figure 7.25 Bearing capacity failure during compaction (pumping clay).

nance. For example, Fig. 7.26 shows the exposed foundation of the Bunker Hill Monument, which was erected at the top of Bunker Hill in 1825 in tribute to the courageous Minutemen who died at Breed's Hill. Figure 7.27 shows the mortar between the hornblende granite blocks that compose the foundation. In some cases the mortar was completely disintegrated, while in other cases, the mortar could be easily penetrated such as shown in Fig. 7.27, where a pen has been used to pierce the mortar. The causes of the disintegration were chemical and physical weathering, such as cycles of freezing and thawing of the foundation.

Because of their age, historic structures tend to be brittle and easily damaged. Feld and Carper (1997) describe the construction of the John Hancock Tower that damaged the adjacent historic Trinity

Figure 7.26   Exposed foundation of the Bunker Hill Monument.

Figure 7.27   Deteriorated mortar between the granite blocks of the Bunker Hill Monument foundation.

Church, in Boston. According to court records, the retaining walls (used for the construction of the John Hancock basement) moved 84 cm (33 in.) as the foundation was under construction in 1969. This movement of the retaining walls caused the adjacent street to sink 46 cm (18 in.) and caused the foundation of the adjacent Trinity Church to shift, which resulted in structural damage and 13 cm (5 in.) tilting of the central tower. The resulting lawsuit was settled in 1984 for about $12 million. An interesting feature of this lawsuit was that the Trinity Church was irreparably damaged, and the damage award was based on the cost of completely demolishing and reconstructing the historic masonry building (ASCE, 1987; *Engineering News-Record,* 1987).

In contrast to Boston where many of the historic structures are made of brick or stone, in the southwestern United States the historic structures are commonly constructed of adobe. Adobe is a sundried brick made of soil mixed with straw. The preferred soil type is a clayey material that shrinks upon drying to form a hard, rocklike brick. Straw is added to the adobe to provide tensile reinforcement. The concept is similar to the addition of steel fibers to modern concrete.

Adobe construction was developed thousands of years ago by the indigenous people, and traditional adobe brick construction was commonly used prior to the twentieth century. The use of adobe as a building material was due in part to the lack of abundant alternative construction materials, such as trees. The arid climate of the southwestern United States also helped to preserve the adobe bricks.

Adobe structures in the form of multifamily dwellings were typically constructed in low-lying valleys, adjacent to streams, or near springs. In an arid environment, a year-round water source was essential for survival. Mud was scooped out of streambeds and used to make the adobe bricks. As a matter of practicality, many structures were constructed close to the source of the mud. These locations were frequently in floodplains, which were periodically flooded, or in areas underlain by a shallow groundwater table.

Adobe construction does not last forever; many historic adobe structures have simply "melted" away. The reasons for deterioration of adobe with time include the uncemented or weakly cemented nature of adobe that makes it susceptible to erosion or disintegration from rainfall, periodic flooding, or water infiltration due to the presence of a shallow groundwater table. For many historic adobe structures, and typical of archeological sites throughout the world, all that remains is a mound of clay.

7.7.1 Case study

The purpose of this section is to describe the performance of a historic adobe structure known as the Guajome Ranch House, located in Vista, California. The Guajome Ranch House, a one-story adobe structure, is considered to be one of the finest large Mexican colonial ranch houses remaining in southern California ("Guajome," 1986). The name Guajome means *frog pond* (Engstrand and Ward, unpublished report, 1991).

The main adobe structure was built during the period 1852–1854. It was constructed with the rooms surrounding and enclosing a main inner courtyard, which is typical Mexican architecture (Fig. 7.28). The main living quarters were situated in the front of the house. Sleeping quarters were located in another wing, and the kitchen and bakehouse in a third wing. At the time of initial construction, the house was sited in a vast rural territory; it was self-sufficient with water from a nearby stream, and the food supply for its habitants was provided by farming and livestock.

More rooms were added around 1855, enclosing a second courtyard (Fig. 7.28). Several more rooms were added in 1887 as indicated in Fig. 7.28. Contemporary additions (not shown in Fig. 7.28) include a garage and gable-roof sewing room (built near the parlor).

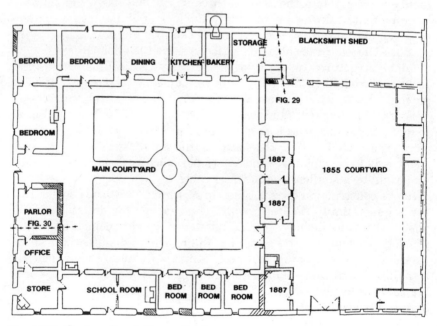

Figure 7.28   Guajome Ranch House site plan.

The Guajome Ranch was purchased by the County of San Diego from the Couts family in 1973. The Guajome Ranch has been declared a National Historic Landmark; restoration was performed by the County of San Diego. The author was a member of the restoration team and investigated the damage caused by poor surface drainage at the site.

Deteriorated condition. The original exterior adobe walls, which are 0.6 to 1.2 m (2 to 4 ft) thick, have been covered with stucco as a twentieth-century modification. Portions of the original low-pitched tile roof have been replaced with corrugated metal.

The Guajome Ranch House was built on gently sloping topography. In the 1855 courtyard, water drained toward one corner and then passed beneath the building. The arrows in Fig. 7.28 indicate the path of the surface water. At the location where the water comes in contact with the adobe, there was considerable deterioration, as shown in Fig. 7.29. Note in this photograph that the individual adobe blocks are visible.

Both the interior and exterior of the structure had considerable deterioration due to a lack of maintenance. The adobe was further eroded when the gable-roofed sewing room burned in 1974 and water from the fire hoses severely eroded the adobe. The hatch walls in Fig. 7.28 indicate those adobe walls having the most severe damage.

Figure 7.29 Deterioration of adobe.

Figure 7.30 Drainage beneath parlor.

The main courtyard drains toward the front of the house, as indicated by the arrows in Fig. 7.28. There are no drains located underneath the front of the house; water is removed by simply letting it flow in ditches beneath the floorboards. Figure 7.30 shows the drainage beneath the front of the house. The moisture has contributed to the deterioration of the wood floorboards. Because the water is in contact with the adobe, the parlor had some of the most badly damaged adobe walls in the structure.

Laboratory testing of original adobe materials. Classification tests performed on remolded samples of the original adobe bricks indicate the soil can be classified as a silty sand to clayey sand (SM-SC), the plasticity index varies from about 2 to 4, and the liquid limit is about 20. On a dry weight basis, the original adobe bricks contain about 50 percent sand-size particles, 40 percent silt-size particles, and 10 percent clay-size particles smaller than 0.002 mm.

The resistance of the original adobe bricks to moisture is an important factor in their preservation. To determine the resistance of the adobe to moisture infiltration, an index test for the erosion potential of an adobe brick was performed (Day, 1990b). The test consisted of trimming an original adobe brick to a diameter of 6.35 cm (2.5 in.)

and a height of 2.54 cm (1.0 in.). Porous stones having a diameter of 6.35 cm (2.5 in.) were placed on the top and bottom of the specimen. The specimen of adobe was then subjected to a vertical stress of 2.9 kPa (60 psf) and was unconfined in the horizontal direction. A dial gauge measured vertical deformation.

After obtaining an initial dial gauge reading, the adobe specimen was submerged in distilled water. Time-versus-dial readings were then recorded. The dial readings were converted to percent strain and plotted versus time, as shown in Fig. 7.31. When initially submerged in distilled water, some soil particles sloughed off in the horizontal (unconfined) direction, but the specimen remained essentially intact. This indicates that there is a weak bond between the soil particles.

After the specimen had been submerged in distilled water for 8 days, the water was removed from the apparatus and the adobe specimen was placed outside and allowed to dry in the summer sun. After 7 days of drying, the adobe specimen showed some shrinkage and corresponding cracking.

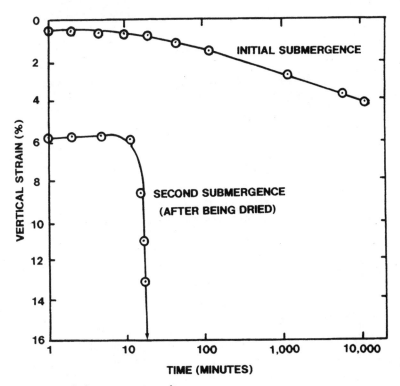

Figure 7.31 Laboratory test results.

The adobe specimen was again submerged in distilled water and after only 18 minutes, the specimen had completely disintegrated as soil particles sloughed off in the horizontal (unconfined) direction. The experiment demonstrated the rapid disintegration of the adobe when it is subjected to wetting-drying cycles. Initially, the adobe is weakly cemented and resistant to submergence, but when the soaked adobe is dried, it shrinks and cracks, allowing for an accelerated deterioration when again submerged.

Drainage repair. Drainage was probably not a major design consideration when the Guajome Ranch House was built. However, the deterioration of the adobe where water is present indicates the importance of proper drainage. The drainage repair consisted of regrading of the courtyards so that surface water flows to box-and-grate inlets connected to storm-water piping. Surface drainage water is removed from the site through underground pipes. The new drainage system prevents surface water from coming in contact with the adobe foundation.

In summary, the interior and exterior of the Guajome Ranch House had severe adobe and wood floor deterioration in areas where water from surface drainage came in contact with the structure (Figs. 7.29 and 7.30). The results of laboratory testing demonstrated the rapid disintegration of the adobe when it is subjected to cycles of wetting and drying. The repair for damage related to surface drainage consisted of a new drainage system that prevents surface water from coming in contact with the adobe foundation.

7.8   Unusual Soil

An unusual soil can be defined as a soil that has rare or unconventional engineering behavior. Such soils can cause damage, distress, or cost overruns (ancillary damages) because their unusual properties were not identified or properly evaluated during the design and construction of a project. Some unusual soils are discussed below:

Rock flour (or bull's liver). This soil consists predominately of silt-size particles, but has little or no plasticity. Nonplastic rock flour contains particles of quartz, ground to a very fine state by the abrasive action of glaciers. Terzaghi and Peck (1967) state that because of its fine particle size, this soil is often mistaken as clay.

Peat. Peat is composed of partially decayed organic matter (humic and nonhumic substances), where the remains of leaves, stems, twigs, and roots can be identified. The places where peat accumulates are known as *peat bogs* or *peat moors*. Its color ranges from light brown to black. Peat is unusual because it has a very high water content, which makes it extremely compressible. This almost always makes it unsuitable for supporting foundations (Terzaghi and Peck, 1967).

Nonwelded tuff and volcanic ash. Tuff is a pyroclastic rock, originating as airborne debris from explosive volcanic eruptions. The largest fragments (in excess of 64 mm) are called *blocks* and *bombs,* fragments between 4 mm and 64 mm are called *lapilli,* fragments between 4 mm and 0.25 mm are *ash,* and the finest fragments (less than 0.25 mm) are *volcanic dust* (Compton, 1962).

An important aspect of tuff is the degree of welding, which can be described as either nonwelded, partially welded to varying degrees, or densely welded. Welding is generally caused by fragments that are hot when deposited, and because of this heat, the sticky glassy fragments may actually fuse together (Best, 1982). There are distinct changes in the original shards and pumice fragments, such as the union and elongation of the glassy shards and flattening of the pumice fragments, which is characteristic of completely welded tuff (Ross and Smith, 1961). The degree of welding depends on many factors; such as type of fragments, plasticity of the fragments (which depends on the emplacement temperature and chemical composition), thickness of the resulting deposit, and rate of cooling (Smith, 1960).

Deposition of volcanic ash directly from the air may result in an unconsolidated (geologically speaking) deposit, which would then be called *ash,* but indurated deposits are called *tuff.* Nonwelded tuff has an engineering behavior similar to volcanic ash. These materials have been used as mineral filler in highways and other earth-rock construction. Some types of volcanic ash have been used as pozzolanic cement and as admixtures in concrete to retard undesired reactions between cement alkalies and aggregates.

Natural deposits of nonwelded tuff and volcanic ash are unusual because they have very low dry densities (e.g., 1 Mg/m^3) due to the presence of lightweight glass and pumice. The materials are also highly susceptible to erosion, which can cause the development of unusual eroded landforms known as *pinnacles.*

Loess. Loess is widespread in the central portion of the United States. It consists of uniform cohesive windblown silt, commonly light brown, yellow, or gray in color, with most of the particle sizes between 0.01 and 0.05 mm (Terzaghi and Peck, 1967). The cohesion is commonly due to calcareous cement which binds the particles together. An unusual feature of loess is the presence of vertical root holes and fractures that make it much more permeable in the vertical direction than the horizontal direction. Another unusual feature of loess is that it can form near vertical slopes, but when saturated, the cohesion is lost and the slope will fail or the ground surface settle.

Caliche. This type of material is common in arid or semiarid parts of the southwestern United States. It consists of soil that is normally cemented together by calcium carbonate. When water evaporates near or at ground surface, the calcium carbonate is deposited in the void spaces between soil particles. Caliche is generally strong and stable in an undisturbed state, but it can become unstable if the cementing agents are leached away by water from leaky pipes or sewers or from the infiltration of irrigation water.

Debris flow and alluvial fan deposits. As previously mentioned, a debris flow can transport a wide variety of soil particle sizes, including boulders and cobbles. Boulders, cobbles, and coarse gravel are typically described as oversize particles and the finer soil particles are described as the soil matrix. Figure 7.32 shows a deposit of oversize particles. The oversize particles can be deposited in alluvial fans or from debris flow, with the finer soil particles filling in the void spaces. The debris flow and alluvial fan deposits often consist of oversize par-

Figure 7.32 Deposit of oversize particles.

ticles that primarily carry the overburden pressure, such as shown in Fig. 7.33, where the four identified cobbles are in direct contact and are carrying the overburden pressure. Such debris flow and alluvial fan deposits are often unstable because of the erratic and unsteady arrangement of oversize particles and loose matrix soil.

Varved clay.   Varved clays ordinarily form as lake deposits and consist of alternating layers of soil. Each varve represents the deposition during a year, with the lower coarse-grained part deposited during the summer, and the upper finer-grained part deposited during the winter when the surface of the lake is frozen and the water is tranquil. This causes an unusual variation in shear strength in the soil, where the horizontal shear strength along the finer-grained (clay) portion of the varve is much less than the vertical shear strength. This can cause the stability of structures founded on varved clay to be overestimated, resulting in a bearing-capacity-type failure.

Bentonite.   Bentonite is a deposit consisting mainly of montmorillonite clay particles. It is derived from the alteration of volcanic tuff or ash. Bentonite is mined to make products that are used as impermeable barriers, such as geosynthetic clay liners (GCLs), which are bentonite/geosynthetic composites. Because bentonite consists almost exclusively of montmorillonite, it will swell, shrink, and cause more expansive soil-related damage than any other type of soil.

Sensitive or quick clays.   The sensitivity of a clay is defined as the undisturbed or natural shear strength divided by its remolded shear strength. Based on this value, the clay can be rated as having a sensitivity from "low" to "quick" (see Table 11.7 in Holtz and Kovacs, 1981). An unusual feature of highly sensitive or quick clays is that the *in situ* moisture content is often greater than the liquid limit (liquidity index greater than 1). Sensitive clays have unstable bonds between particles. As long as these unstable bonds are not broken, the clay can support a heavy load. But once remolded, the bonding is destroyed and the shear strength is substantially reduced. For example, sensitive Leda Clay, from Ottawa, Ontario, has a high shear strength in the undisturbed state, but once remolded, the clay is essentially a fluid (no shear strength). There are reports of entire hillsides of sensitive or quick clays becoming unstable and then simply flowing away (Lambe and Whitman, 1969).

Diatomaceous earth.   Diatoms are defined as microscopic, single-celled plants of the class Bacillariophyceae, which grow in both marine and fresh water (Bates and Jackson, 1980). Diatoms secrete outer shells of

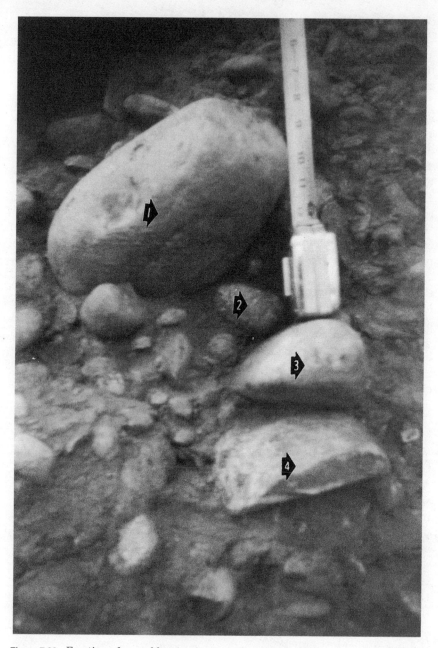

Figure 7.33 Erratic and unstable arrangement of oversize particles.

silica, called frustules, in a great variety of forms which can accumulate in sediments in enormous amounts (Bates and Jackson, 1980). Deposits of diatoms have low dry density and high moisture content because the structure of the diatom is an outer shell of silica that can contain water. Common shapes of diatoms are rodlike, spherical, or circular disks having a typical length or diameter of about 0.03 to 0.11 mm (Fig. 7.34, from Spencer, 1972). Diatoms typically have rough surface features, such as protrusions or indentations.

A natural deposit of diatoms is commonly referred to as *diatomaceous earth* or *diatomite*. Diatomaceous earth usually consists of fine, white, siliceous powder, composed mainly of diatoms or their remains (Terzaghi and Peck, 1967; Stokes and Varnes, 1955). Diatomite is a organogenetic sedimentary rock containing frustules of diatoms and

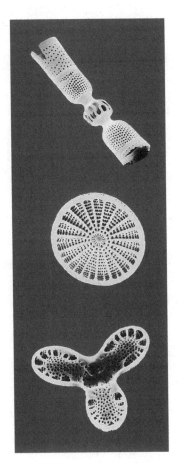

Figure 7.34 Diatoms; top diatom enlarged approximately 1200 times, middle and bottom diatoms enlarged approximately 300 times. (*Reprinted from The Dynamics of the Earth, Copyright 1972 by the Thomas Y. Crowell Company, Inc.*)

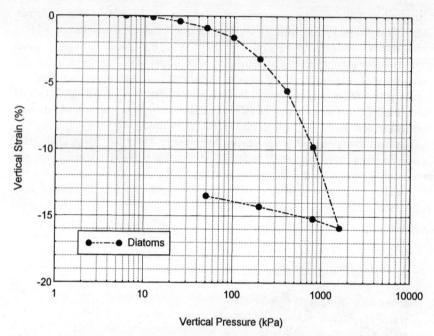

Figure 7.35 Vertical strain versus vertical pressure [*From Day (1995c), reprinted with permission from the American Society of Civil Engineers.*]

sometimes mixed with shells of radiolarians, spicules of sponges, and foraminifera (Mottana et al., 1978). Industrial uses of diatomaceous earth or diatomite are as filters to remove impurities, as abrasives to polish soft metals, and when mixed with nitroglycerin, as an absorbent in the production of dynamite (Mottana et al., 1978).

Diatomaceous earth can be very compressible when used as fill. For example, Fig. 7.35 shows the one-dimensional vertical settlement of diatomaceous earth measured by the oedometer apparatus. The initial dry density of the diatomaceous earth was 0.87 Mg/m^3 (54.3 pcf). During the testing of the diatomaceous earth, there was a distinct popping sound at high vertical pressure (i.e., 1600 kPa) which was due to the diatoms (which are essentially hollow shells of silica) being crushed together. The crushing together of the diatoms is the reason for the high compressibility at high vertical pressures.

7.8.1 Case study

This section presents a case study of the deformation of fill slopes at Canyon Estates, located in Mission Viejo, California. The author was retained as an expert for the plaintiffs. The case settled out of court in early 1997 for over $10 million. The purpose of this case study is to describe the slope movement, present data on observed damages, and discuss the causes of the slope movement. A discussion of the legal issues of the project, which are important for geotechnical engineers practicing in the United States, will also be presented. As of the date of preparation of this case study, no repairs had been performed at the site.

The Canyon Estates project is an approximately 350-acre parcel of land. Earth moving operations were utilized to fill canyons in order to develop roads and terraced level building pads in 1983 to 1988. Approximately 750 one- and two-story single-family residences had been built on these pads. At the time of the study, there were an additional 200 houses in various stages of construction. Structures appurtenant to all the homes include side yard property line block walls which are typically connected to top-of-slope iron post and block walls. Homeowner improvements include patio flatwork, decks, and pools.

Because of the filled-in canyons and buttressed slopes, most building pads contained rear-yard descending fill slopes with some building pads also containing a side-yard descending fill slope. The fill slopes were constructed at approximately 2:1 (horizontal:vertical) inclinations and commonly range between 6 to 12 m (20 to 40 ft) in height throughout the site. In some areas, the fill slopes were constructed to approximately 20 m (65 ft) in height. Because of the presence of adverse (out of slope) bedding in cut slope areas, the cut slopes were overexcavated and buttressed with fill. These buttressed slopes were generally constructed with an equipment-width (4.6 m, 15 ft) key and extended from the slope toe to the top.

Monterey formation.   The fill used to create the fill slopes at Canyon Estates was derived from the Monterey formation. The Monterey formation is generally classified as a siltstone or claystone. An unusual feature of the Monterey formation is that it contains diatoms or broken fragments of diatoms. As previously discussed, deposits of diatoms are unusual because they have a low dry density and high moisture content due to the structure of the diatom which is an outer shell of silica that can contain water (Fig. 7.34).

Index properties of fill.   Using the Unified Soil Classification System, the fill at Canyon Estates was classified as a silty clay (CH) to clayey silt (MH). The liquid limit was generally between 60 to 70. Based on 24 particle-size analyses of fill throughout the site, the percentage of clay particles (finer than 0.002 mm) varied from 28 percent to 52 percent, based on dry weight. X-ray diffraction tests indicated that the clay particles are predominantly montmorillonite. A majority of the remaining particles are of silt size, with a significant portion being the remains of diatoms. The diatoms do affect the index properties because in some cases the limits plot well below the A line, while inorganic soil containing montmorillonite usually plots just below the U line (see Fig. 4.14 in Holtz and Kovacs, 1981). In some areas, the process of compaction did not completely pulverize the fragments of Monterey formation that were ripped from cut areas, and the fill did contain oversize particles, but this was usually less than 10 percent, based on dry weight.

New method of compaction (ADC). Because of the high moisture content of the diatomaceous material encountered during site grading and the difficulties in compacting the soil to the industry standard of a minimum of 90 percent relative compaction (based on Modified Proctor), an alternative method of determining fill compaction was utilized. This alternative method required compacting the fill to 95 percent of the maximum achievable density of compaction (ADC). The maximum achievable density is the soil density obtained by compacting a sample (one point) at an *in situ* moisture content using the Modified Proctor compaction energy. In essence, the unusual nature of the diatomaceous earth resulted in the development of a new method for the placement and testing of the soil at the site.

As a result of using the ADC test method, fill was compacted at a higher than optimum moisture content and to a relative compaction that was less than the industry standard. Average relative compaction measured during the investigation was about 82 percent. Because of the lower density achieved by the ADC method, there was a corresponding reduction in shear strength which allowed the slopes to deform more than typically expected. For example, Figs. 7.36 through 7.38 present unconsolidated-undrained (UU) triaxial compression tests performed on typical fill specimens remolded to relative compaction of either 70 percent (Fig. 7.36), 80 percent (Fig. 7.37), or 90 percent (Fig. 7.38) and tested at different confining pressures. This data shows a substantial reduction in the undrained shear strength as the relative compaction decreases.

Basis for filing of lawsuit. The complaint was initially filed by the Canyon Estates Homeowners Association in Orange County Superior Court on April 3, 1992. The first and preeminent basis for filing the lawsuit was "strict liability." As mentioned in Sec. 1.4, the concept of strict liability means that developers of mass-produced housing will

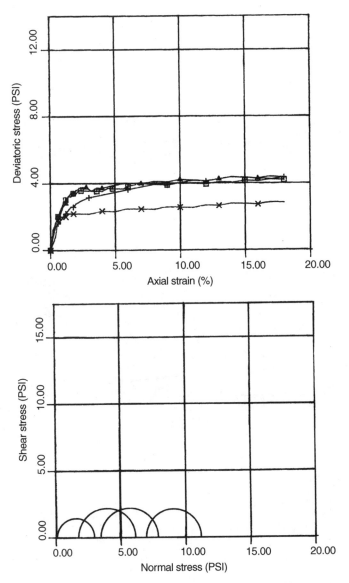

Figure 7.36 UU triaxial tests (relative compaction = 70 percent, moisture content = 49 percent).

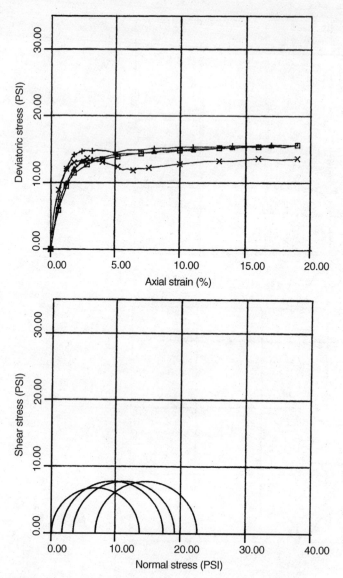

Figure 7.37 UU triaxial tests (relative compaction = 80 percent, moisture content = 39 percent).

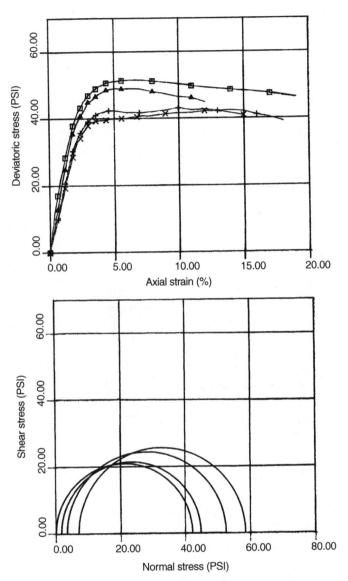

Figure 7.38 UU triaxial tests (relative compaction = 90 percent, moisture content = 28 percent).

be found liable for any defects at the project, regardless of whether the defect resulted from a failure to comply with the standard of practice. In California, it is not necessary to prove negligence on the part of the builder, but rather all that needs to be shown is that the project has a defect. A defect exists when a product or some component of it fails to perform normally when used for its intended purpose by the consumer.

At Canyon Estates, it was contended that the movement of the slopes with consequential failure of lateral support to hundreds of homes atop those slopes was a defect, in that the slopes were failing to perform normally when used for their intended purpose by the consumer.

Damage at the site. Visual inspections of fill slope areas and top-of-slope improvements such as walls, flatwork, and pools revealed a variety of distress. Typically, damage consisted of cracking and separation of block walls near the slope tops. In most cases, the pilasters at the slope top pulled away from the side-yard property line walls (Figs. 7.39 and 7.40). Some of these separations were on the order of 20 cm (8 in.) in width. Another common type of damage to the side-yard property walls was stair-step cracking which occurred up to 4.5 m to 6 m (15 to 20 ft) back from the slope top (Figs. 7.41 and 7.42). Often, cracks extended through the wall footings as shown in Fig. 7.43. The damage indicated predominately a lateral component of movement as shown in Figs. 7.39 to 7.43.

In general, the higher the slope, the greater the observed magnitude of cracking and separations. Also, at lot corners where slopes descended in both the back and side yards, distress was greater. Depending on the type of wall and width of cracking, most damage varied from "architectural" to "functional."

Observations of backyard improvements, such as patio slabs, planter walls, and pool areas have exhibited similar cracks and separations (Figs. 7.44 and 7.45). These features generally align parallel to the slope and in some areas were noted as far as 6 m (20 ft) back from the slope tops, indicating fairly extensive slope movement. In some cases, it was observed that a gap opened up between the concrete driveway and the concrete garage slab. In some cases, the gap exceeded 2.5 cm (1 in.) and the gap was believed to be due to the effect of slope deformation which was pulling the posttensioned house foundations downslope.

During the investigation of site conditions, it was noted that some repairs had been made to the damage. Repairs were noted to consist of localized patching with grout or extending of wrought iron fences to connect pilaster/wall separations. Both types of repair were observed

Figure 7.39 Eight inches of separation of top-of-slope wall.

Figure 7.40 Four inches of separation of top-of-slope wall.

to be ineffective. In most cases, the patches reopened and wrought iron extensions disconnected, indicating ongoing slope movement. Some of the reopened cracks exceeded 5 cm (2 in.).

Inclinometer monitoring. Thirteen inclinometer casings were installed in the rear yards (near the top of slope) throughout the Canyon Estates site in order to monitor the rate and magnitude of lateral movement. The 13 inclinometers were monitored from mid-1994 up to the time the case settled out of court (early 1997). About one-half of the inclinometers displayed inconclusive or low-level slope movement during the monitoring period. The other one-half did show continuous lateral movement during the monitoring period. Those inclinometers

Figure 7.41 Cracking and separation of wall near top of slope.

displaying ongoing movement were generally located at the highest fill slopes or in areas where there were both descending rear- and side-yard slopes.

Figures 7.46 to 7.49 show the downslope (A direction) movement of two of the inclinometers at Canyon Estates using the Geo-Slope inclinometer plot program. In each figure, the plot on the left is the lateral (downslope) deformation versus depth, while the plot on the right is the lateral deformation versus time.

Figures 7.46 to 7.48 are the plots for one of the inclinometers. The right side of Fig. 7.46 shows the lateral deformation versus time for data at a depth of 0.6 m (2 ft), Fig. 7.47 shows the lateral deformation versus time for data at a depth of 1.8 m (6 ft), and Fig. 7.48 shows the lateral deformation versus time for data at a depth of 3.6 m (12 ft).

Figure 7.42 Stair-step cracking of wall near top of slope.

Figure 7.43 Cracking through wall and footing.

Figure 7.44 Lateral movement of pool deck.

Note the cyclic behavior of the lateral deformation versus time for the data at a depth of 0.6 m (2 ft). Since the inclinometer was not installed in the slope, but rather on the flat part of the building pad near the top of slope, this cyclic inclinometer behavior was attributed to wetting and drying cycles. When the building pad dried out, there was a pulling of the inclinometer in the direction away from the slope face. Note that the plots of lateral deformation versus time at depths of 1.8 m (6 ft, Fig. 7.47) and 3.6 m (12 ft, Fig. 7.48) do not show the cyclic lateral deformation, indicating that this phenomenon is a near-surface condition.

Figure 7.49 presents the plot for a second inclinometer. Note in this figure that the depth of movement was rather deep, on the order of 4.9 to 5.5 m (16 to 18 ft). Also, the inclinometers did not indicate a discrete plane of failure, but rather a progressive decrease in lateral movement with depth.

Slope stability. Slope stability total stress (undrained) analysis were performed by using the SLOPE/W (Geo-Slope, 1991) computer program. By inputting the slope configuration (2:1 horizontal:vertical slope inclination), undrained shear strength parameters from Figs. 7.36 to 7.38, and wet density of soil [1.84 Mg/m^3 (115 pcf)], the computer program calculated the factor of safety using the Janbu simpli-

Figure 7.45   Separation between patio and pool bond beam.

fied method of slices. The slope stability analyses are summarized in Table 7.2.

The slope stability analyses show that at a relative compaction of 70 percent, the fill slopes would fail. At 80 percent relative compaction, which is about the average for the site, the taller slopes (18 m, 60 ft) have a low factor of safety of 1.17. Figure 7.50 presents the

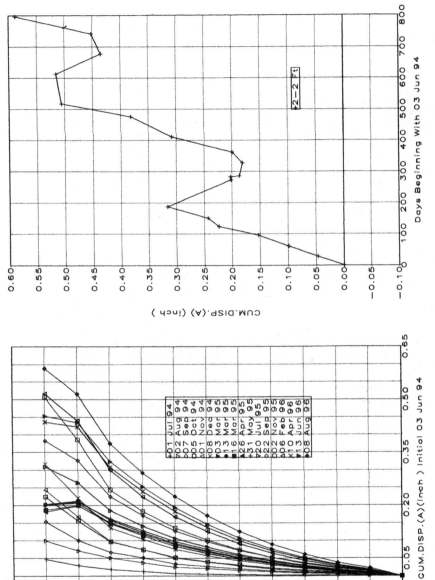

Figure 7.46 Inclinometer 7 (displacement plot for depth = 2 ft).

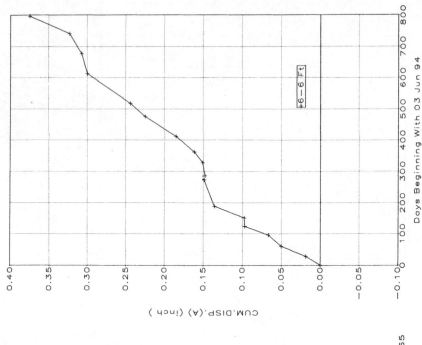

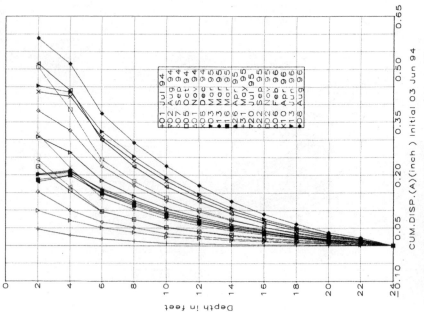

Figure 7.47 Inclinometer 7 (displacement plot for depth = 6 ft).

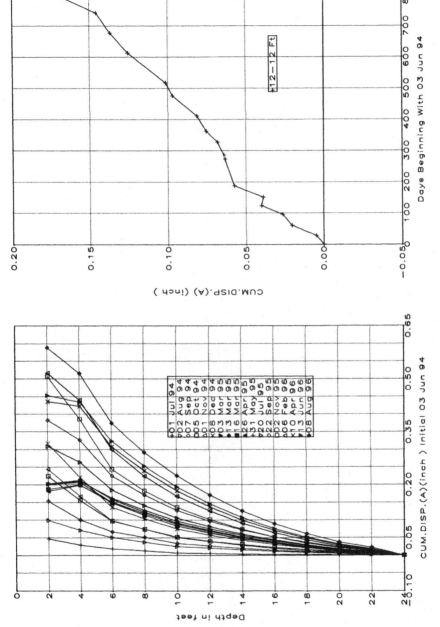

Figure 7.48 Inclinometer 7 (displacement plot for depth = 12 ft).

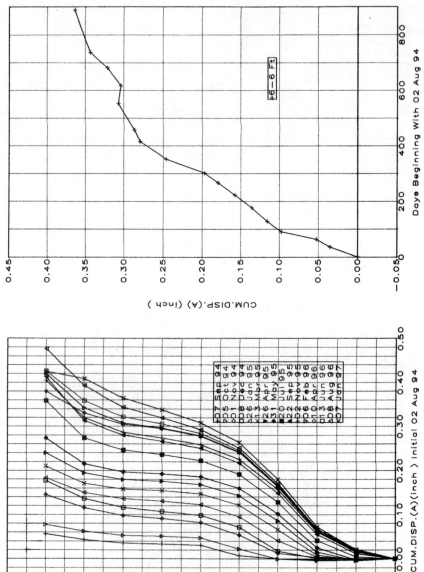

Figure 7.49 Inclinometer 12 (displacement plot for depth = 6 ft).

TABLE 7.2   Summary of Slope Stability Analysis

Relative compaction, % (1)	Undrained shear strength (2)	Slope height, m (ft) (3)	Factor of safety (4)
70	Fig. 7.36	9 (30)	1.15
		18 (60)	0.97
80	Fig. 7.37	9 (30)	1.89
		18 (60)	1.17
90	Fig. 7.38	9 (30)	5.11
		18 (60)	3.65

NOTE: All slope inclinations are 2:1 (horizontal:vertical) and the wet density = 1.84 Mg/m³ (115 pcf).

slope stability analysis for the condition of 80 percent relative compaction for a slope of 18 m (60 ft) height. For a condition of 90 percent relative compaction, the slopes have a very high factor of safety because of the high undrained shear strength.

Slope deformation. The results of the slope stability analyses show a low factor of safety for total stress (undrained) conditions for the higher fill slopes (Table 7.2). From these results, one cause of the lateral movement at the site was determined to be undrained creep of the compacted slopes. The slope stability analyses show that the taller slopes would be most susceptible to the undrained creep.

A second factor in the movement was slope softening. According to a comparison of initial as-compacted moisture contents and those obtained from this investigation, some of the fill slopes were subjected to an increase in moisture content with time. It was concluded that a second probable cause of the slope deformation was slope softening.

As previously mentioned, the near-surface wetting and drying of the building pad did affect the inclinometer (Fig. 7.46). It was concluded that a third probable cause of the near-surface slope deformation was seasonal moisture changes.

Summary. Figures 7.39 to 7.43 show pictures of typical severe damage at the Canyon Estates project. The cause of the damage was lateral deformation of the fill slopes. The data indicated that there were three separate mechanisms that caused the slope deformation: undrained creep, slope softening, and near-surface downslope movement due to seasonal moisture changes.

The results of slope stability analysis (Table 7.2) indicate the importance of relative compaction on the undrained shear strength of

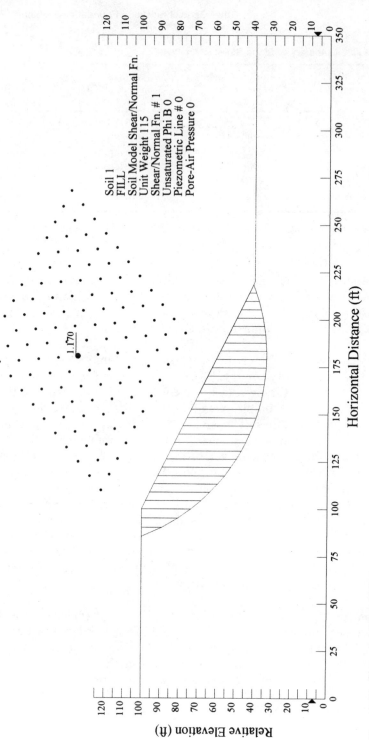

Soil 1
FILL
Soil Model Shear/Normal Fn.
Unit Weight 115
Shear/Normal Fn. # 1
Unsaturated Phi B 0
Piezometric Line # 0
Pore-Air Pressure 0

1.170

Horizontal Distance (ft)

Relative Elevation (ft)

Figure 7.50 Slope stability analysis.

the soil. The project compaction specifications were based on the presence of unusual soil (diatomaceous earth) at the site. As a consequence, a new method of compaction (ADC method) was developed for the site. This compaction method resulted in an average degree of fill compaction of only 82 percent. This low compaction was considered a primary reason for the lateral deformation of the fill slopes.

Engineers must be aware of legal issues when working on specific projects. In California, the courts have adopted the concept of "strict liability." This means that in order to obtain a monetary judgment, negligence need not be proven. All that needs to be shown is that the project has a defect that is causing damage. At the Canyon Estates project, the defect was the lateral deformation of fill slopes that damaged block walls and homeowner improvements.

7.9 Retaining Walls

A retaining wall is defined as a structure whose primary purpose is to provide lateral support for soil or rock. In some cases, the retaining wall may also support vertical loads. Examples include basement walls and certain types of bridge abutments.

Cernica (1995a) lists and describes varies types of retaining walls. Some of the more common types of retaining walls are gravity walls, counterfort walls, cantilevered walls, and crib walls. Gravity retaining walls are routinely built of plain concrete or stone and the wall depends primarily on its massive weight to resist failure from overturning and sliding. Counterfort walls consist of a footing, a wall stem, and intermittent vertical ribs (called counterforts) which tie the footing and wall stem together. Crib walls consist of interlocking concrete members that form cells which are then filled with compacted soil.

Although reinforced earth retaining walls have become more popular in the past decade, cantilever retaining walls are still probably the most common type of retaining structure. There are many different types of cantilevered walls, with the common features being a footing that supports the vertical wall stem. Typical cantilevered walls are T-shaped, L-shaped, or reverse L-shaped (Cernica, 1995a).

To prevent the buildup of hydrostatic water pressure on the retaining wall, clean granular material (no silt or clay) is the standard recommendation for backfill material. Import granular backfill generally has a more predictable behavior in terms of earth pressure exerted on the wall. A backdrain system is often constructed at the heel of the wall to intercept and dispose of any water seepage in the granular backfill.

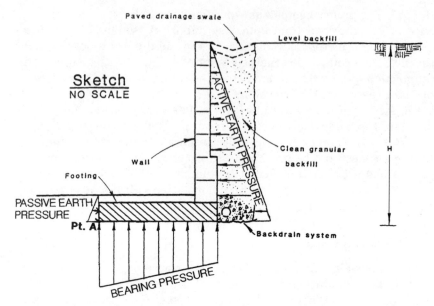

Figure 7.51   Retaining wall design pressures. [*From Day (1997c), reprinted with permission from the American Society of Civil Engineers.*]

Figure 7.51 shows a reverse L-shaped cantilever retaining wall. The pressure exerted on the wall is the active earth pressure. The footing is supported by the vertical bearing pressure of the soil or rock. Lateral movement of the wall is resisted by passive earth pressure and slide friction between the footing and bearing material. The possible failure modes for a retaining wall are discussed below.

Lateral movement.   As shown in Fig. 7.51, the active earth pressure is often assumed to be horizontal by neglecting the friction developed between the vertical wall stem and the backfill. This friction force has a stabilizing effect on the wall and therefore it is usually a safe assumption to ignore friction. However, if the wall should settle more than the backfill—for example, because of high vertical loads imposed on the top of the wall—then a negative skin friction can develop between the wall and backfill which has a destabilizing effect on the wall.

In the evaluation of the active earth pressure, it is common for the soil engineer to recommend clean granular soil as backfill material. In order to calculate the active earth pressure resultant force P_a, in kilonewtons per linear meter of wall or pounds per linear foot of wall, the following equation is used for clean granular backfill:

$$P_a = \tfrac{1}{2}k_a\gamma_t H^2 \tag{7.1}$$

where k_a = active earth pressure coefficient, γ_t = wet density of the backfill, and H = height over which the active earth pressure acts as defined in Fig. 7.51. The active earth pressure coefficient k_a is equal to

$$k_a = \tan^2(45° - \tfrac{1}{2}\phi) \tag{7.2}$$

where ϕ = friction angle of the clean granular backfill.

In Eq. (7.1), the product of k_a times γ_t is referred to as the *equivalent fluid pressure* (even though the product is actually a density). In the design analysis, the soil engineer usually assumes a wet density γ_t of 1.9 Mg/m³ (120 pcf) and a friction angle ϕ of 30° for the granular backfill. Using Eq. (7.2) and $\phi = 30°$, we find the active earth pressure coefficient k_a is 0.333. Multiplying 0.333 by the wet density γ_t of backfill results in an equivalent fluid pressure of 0.64 Mg/m³ (40 pcf).

This is the typical recommendation for equivalent fluid pressure from soil engineers. It is valid for the conditions of clean granular backfill, a level ground surface behind the wall, a backdrain system, and no surcharge loads. Note that this commonly recommended value of equivalent fluid pressure of 0.64 Mg/m³ (40 pcf) does not include a factor of safety and is the actual pressure that would be exerted on a smooth wall when the friction angle ϕ of the granular backfill equals 30°. In designing the vertical wall stem in terms of wall thickness and size and location of steel reinforcement, a factor of safety F can be applied to the active earth pressure in Eq. (7.1). A factor of safety may be prudent because higher wall pressures will most likely be generated during compaction of the backfill or when translation of the footing is restricted (Goh, 1993).

For the case of an inclined slope behind the retaining wall, equations and tables have been developed to determine the active earth pressure coefficient k_a, such as NAVFAC DM 7.2 (1982), page 7.2-64. If there is a uniform surcharge pressure Q acting upon the ground surface behind the wall, then there would be an additional horizontal pressure exerted upon the retaining wall equal to the product of k_a times Q.

In order to develop passive pressure, the wall footing must move laterally into the soil. As indicated in Table 7.3 (from NAVFAC DM 7.2, 1982), the wall translation to reach the passive state is at least twice that required to reach the active earth pressure state.

Usually it is desirable to limit the amount of wall translation by applying a reduction factor to the passive pressure. A commonly used reduction factor is 2.0 (Lambe and Whitman, 1969). The soil engineer

TABLE 7.3   Magnitudes of Wall Rotation to Reach Failure

Soil type and condition (1)	Rotation (Y/H) for active state (2)	Rotation (Y/H) for passive state (3)
Dense cohesionless	0.0005	0.002
Loose cohesionless	0.002	0.006
Stiff cohesive	0.01	0.02
Soft cohesive	0.02	0.04

NOTE: Y = wall displacement and H = height of the wall.
SOURCE: NAVFAC DM-7.2, 1982.

routinely reduces the passive pressure by half (reduction factor = 2.0) and then refers to the value as the *allowable passive pressure.* To limit wall translation, the structural engineer should use the allowable passive pressure for design of the retaining wall.

The passive pressure may also be limited by building codes. For example, the allowable passive soil pressure, in terms of equivalent fluid pressure, is 16 to 32 kN/m^3 (100 to 200 pcf) per the *Uniform Building Code* (1997).

Bearing capacity failure.   In order to calculate the footing bearing pressure, the first step is to sum the vertical loads, such as the wall and footing weights. The vertical loads can be represented by a single resultant vertical force W, per linear meter or foot of wall, that is offset by a distance x' from the toe of the footing. The resultant force W and the distance x' can then be converted to a pressure distribution as shown in Fig. 7.51 (see Lambe and Whitman, 1969, Example 13.12). The largest bearing pressure is routinely at the toe of the footing (Point A, Fig. 7.51). The largest bearing pressure should not exceed the allowable bearing pressure, which is usually provided by the soil engineer or by local building code specifications.

Sliding failure.   The factor of safety F for sliding of the retaining wall is defined as the resisting forces divided by the driving force (the forces are per linear meter or foot of wall):

$$F = \frac{\text{sliding friction force} + \text{allowable passive resultant force}}{\text{active earth pressure resultant force}}$$

$$= \frac{\mu W + P_p}{P_a} \tag{7.3}$$

where μ = friction coefficient between the concrete foundation and bearing soil, W = resultant vertical force, P_p = passive resultant force, and P_a = active earth resultant force calculated from Eq. (7.1). The typical recommendations for minimum factor of safety for sliding is 1.5 to 2.0 (Cernica, 1995a).

In some situations, there may be adhesion between the bottom of the footing and the bearing soil. This adhesion is often neglected because the wall is designed for active pressures, which typically develop when there is translation of the footing. Translation of the footing will break the adhesive forces between the bottom of the footing and the bearing soil and therefore adhesion is often neglected for the factor of safety of sliding.

Overturning failure. The factor of safety F for overturning of the retaining wall is calculated by taking moments about the toe of the footing (point A, Fig. 7.51), and is

$$F = \frac{\text{stabilizing moment}}{\text{overturning moment}} = \frac{Wx'}{\frac{1}{3}P_a H} \qquad (7.4)$$

where x' = distance from the resultant vertical force W to the toe of the footing, and P_a = active earth resultant force calculated from Eq. (7.1). The typical recommendations for minimum factor of safety for overturning is 1.5 to 2.0 (Cernica, 1995a).

Common causes of failure. There are many different reasons for excessive lateral movement, bearing capacity failures, sliding failures, or failure by overturning of the retaining wall. Common causes include inadequate design, improper construction, or unanticipated loadings. Other causes of failure are listed below:

1. *Clay backfill.* A frequent cause of failure is that the wall was backfilled with clay. As previously mentioned, clean granular sand or gravel is usually recommended as backfill material. This is because of the undesirable effects of using clay or silt as a backfill material (Sec. 5.3.1). When clay is used as backfill material, the clay backfill can exert swelling pressures on the wall (Fourie, 1989; Marsh and Walsh, 1996). The highest swelling pressures develop when water infiltrates a backfill consisting of a clay that was compacted to a high dry density at a low moisture content. The type of clay particles that will exert the highest swelling pressures are those of montmorillonite. Because the clay backfill is not free-draining, there could also be additional hydrostatic forces or ice-related forces that substantially increase the thrust on the wall.

2. *Inferior backfill soil.* To reduce construction costs, soil available on site is sometimes used for backfill. This soil may not have the properties, such as being a clean granular soil with a high shear strength, that was assumed during the design stage. Using on-site available soil, rather than imported granular material, is probably the most common reason for retaining wall failures.

3. *Compaction-induced pressures.* As previously mentioned, one reason for applying a factor of safety F to the active earth pressure [Eq. (7.1)] is because larger wall pressures will typically be generated during compaction of the backfill. By using heavy compaction equipment in close proximity to the wall, excessive pressures can be developed that damage the wall. The best compaction equipment, in terms of exerting the least compaction induced pressures on the wall, are small vibrator plate (hand-operated) compactors such as models VPG 160B and BP 19/75 (Duncan et al., 1991). The vibrator plates effectively densify the granular backfill, but do not induce high lateral loads because of their light weight. Besides hand-operated compactors, other types of relatively lightweight equipment can be used to compact the backfill. For example, Fig. 7.52 shows a bobcat being used to place and compact the backfill.

4. *Failure of the backcut.* There could also be the failure of the backcut for the retaining wall. The vertical backcut shown in Fig. 7.51 is often used when the retaining wall is less than 1.5 m (5 ft) high. In other cases, the backcut is usually sloped. Figure 7.53 shows an example of the back-cut slope for the construction of a retaining wall in San Carlos, California. The backcut can fail if it is excavated too steeply and does not have an adequate factor of safety.

7.9.1   Case study

The case study involves a retaining wall failure in San Diego, California. The wall was constructed as a basement wall for a large building. In 1984, the building was demolished and the site was turned into a parking lot.

As originally constructed, the basement wall received lateral support from the foundation, a bowstring roof truss, and perpendicular building walls. When the building was demolished, the retaining wall essentially became a cantilevered wall with no lateral support except from the footing.

The retaining wall is about 2.4 to 2.7 m (8 to 9 ft) high, 20 cm (8 in.) thick, with thickened pilasters that originally supported the bowstring roof truss. Figure 7.54 shows a photograph of the wall after demolition of the building. The area behind the wall belonged to an adjacent property owner, who experienced damage when the wall

Figure 7.52 Compaction of backfill. [*From Day (1997c), reprinted with permission from the American Society of Civil Engineers.*]

moved due to the loss of lateral support. Figure 7.55 shows a photograph of cracks that opened up in the concrete flatwork located behind the retaining wall.

The movement of the wall was monitored by installing brass pins on opposite sides of the flatwork cracks. By measuring the distance between the pins, the opening of the cracks (lateral movement) was calculated and plotted versus time as shown in Fig. 7.56. The horizontal axis in Fig. 7.56 is time after installation of the crack pins.

Note in Fig. 7.56 that the movement of the wall versus time is not at a constant rate, but rather intermittent. The data indicates that the wall moves forward, the cracks open up, and then lateral movement ceases for awhile. This is because the soil thrust is reduced

Figure 7.53 Backcut slope for retaining wall construction. [*From Day (1997c), reprinted with permission from the American Society of Civil Engineers.*]

Figure 7.54 Cantilevered retaining wall. [*From Day (1997c), reprinted with permission from the American Society of Civil Engineers.*]

Figure 7.55 Area behind retaining wall (arrows point to cracks opening up in the flatwork). [*From Day (1997c), reprinted with permission from the American Society of Civil Engineers.*]

when the wall moves forward, and it takes time for the soil to reassume its original contact with the back face of the wall. In Fig. 7.56, crack pin (CP) 3 did record a closing of the crack at a time of 0.9 to 1.2 years, but this is due to settlement of the backfill and flatwork as the soil reassumed contact with the back face of the wall. Figure 7.57 shows the voids that developed beneath the flatwork due to lateral movement of the wall.

As illustrated by this case study, most retaining wall failures are gradual and the wall slowly fails by intermittently tilting or moving laterally. It is possible that a failure can occur suddenly, such as when there is a slope-type failure beneath the wall or when the foundation

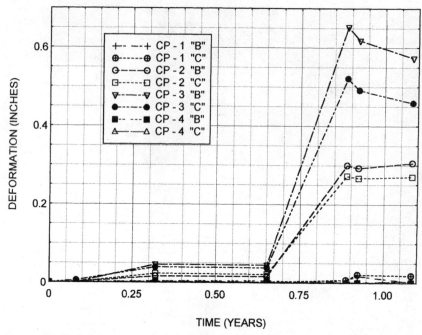

Figure 7.56 Wall deformation versus time. [*From Day (1997c), reprinted with permission from the American Society of Civil Engineers.*]

of the wall fails due to inadequate bearing capacity. These rapid failure conditions could develop if the wall foundation is supported by clay (Cernica, 1995a).

Another example of sudden wall failure could be during an earthquake. It is difficult to accurately predict the additional lateral forces that will be generated on a retaining wall during an earthquake. Some factors affecting the magnitude of earthquake forces on the wall are the size and duration of the earthquake, the distance from the earthquake epicenter to the site, and the mass of soil retained by the wall. Many retaining walls are designed for only the active earth pressure and then fail when additional forces are generated by the earthquake.

Figure 7.57 Voids that developed below flatwork as a result of wall movement. [*From Day (1997c), reprinted with permission from the American Society of Civil Engineers.*]

8

Groundwater and Moisture Problems

The following notation is introduced in this chapter:

Symbol	Definition
A	Surface area of the moisture dome test
Q	Weight of absorbed water from the moisture dome test
T	Time in days that the moisture dome test was performed
V	Vapor flow rate or emission rate

8.1 Groundwater

Groundwater can cause or contribute to failure because of excess saturation, seepage pressures, or uplift forces. It has been stated that uncontrolled saturation and seepage causes many billions of dollars a year in damage (Cedergren, 1989). Some of the more common problems due to groundwater are as follows (Harr, 1962; Collins and Johnson, 1988; Cedergren, 1989):

- Piping failures of dams, levees, and reservoirs

- Seepage pressures that cause or contribute to slope failures

- Deterioration and failure of roads due to the presence of groundwater in the base or subgrade

- Highway and other fill foundation failures caused by perched groundwater

- Earth embankment and foundation failures caused by excess pore water pressures

- Retaining wall failures caused by hydrostatic water pressures
- Canal linings, drydocks, and basement or spillway slabs uplifted by groundwater pressures
- Soil liquefaction, cause by earthquake shocks, because of the presence of loose granular soil that is below the groundwater table
- Transportation of contaminants by the groundwater

8.1.1 Pavements

Probably the most common engineering facilities damaged by groundwater are pavements. It has been stated that groundwater in pavements accelerates the damage rates by hundreds of times over the damage rates of pavements with no groundwater. This premature failure of thousands of miles of pavements and billions of dollars in losses a year could be avoided by good pavement drainage practices (Cedergren, 1989). The key element in a good drainage system is a layer of highly permeable material (such as open graded gravel) protected by filters or geofrabric so that the permeable material will not become clogged by the intrusion of soil fines. A drainage system, to remove the water from the base, is also required.

There are several ways that groundwater can enter the base material. In areas having a high groundwater table or artesian condition, water can be forced upward into the base material. Water can also flow downward through pavement cracks or joints. There can also be the development of a perched groundwater condition, where water moves laterally through the base from adjacent planter areas, medians, or shoulders.

Figure 8.1 shows an example of the effect of groundwater on pavement deterioration. At this site, water was observed to be coming up through the pavement as shown in Fig. 8.1. A test pit was excavated in a utility trench, with the result that groundwater bubbled up from the ground as shown in Fig. 8.2. It was observed that the native soil was clayey, but the utility trenches had been backfilled with granular (permeable) soil. At the low points in the streets, the water exited the utility trenches and flowed up through the pavement surface as shown in Fig. 8.1. The continuous flow of groundwater from the utility trenches led to premature deterioration of the pavements.

One method to prevent the flow of groundwater through utility trenches is to use a grout (such as a cement slurry) to encase the utility lines. The remainder of the trench could then be backfilled and compacted with on-site native soil. This should provide the trench with a permeability equal to or less than the surrounding native soil.

Figure 8.1   Flow of groundwater through top of pavement.

Figure 8.2   Groundwater exiting a test pit excavated into a utility trench.

8.1.2 Slopes

Groundwater can affect slopes in different ways. Table 8.1 presents common examples and the influence of groundwater on slope failures. The main destabilizing factors of groundwater on slope stability are as follows (Cedergren, 1989):

1. Reducing or eliminating cohesive strength
2. Producing pore water pressures which reduce effective stresses, thereby lowering shear strength
3. Causing horizontally inclined seepage forces which increase the driving forces and reduce the factor of safety of the slope

TABLE 8.1   Common Examples of Slope Failures

Kind of slope (1)	Conditions leading to failure (2)	Type of failure and its consequences (3)
Natural earth slopes above developed land areas (homes, industrial)	Earthquake shocks, heavy rains, snow, freezing and thawing, undercutting at toe, mining excavations	Mud flows, avalanches, land-slides; destroying property, burying villages, damming rivers
Natural earth slopes within developed land areas	Undercutting of slopes, heaping fill on unstable slopes, leaky sewers and water lines, lawn sprinkling	Usually slow creep type of failure; breaking water mains, sewers, destroying buildings, roads
Reservoir slopes	Increased soil and rock saturation, raised water table, increased buoyancy, rapid drawdown	Rapid or slow landslides, damaging highways, railways, blocking spillways, leading to overtopping of dams, causing flood damage with serious loss of life
Highway or railway cut or fill slopes	Excessive rain, snow, freezing, thawing, heaping fill on unstable slopes, undercutting, trapping groundwater	Cut slope failures blocking roadways, foundation slipouts removing roadbeds or tracks, property damage, some loss of life
Earth dams and levees, reservoir ridges	High seepage levels, earthquake shocks; poor drainage	Sudden slumps leading to total failure and floods downstream, much loss of life, property damage
Excavations	High groundwater level, insufficient groundwater control, breakdown of dewatering systems	Slope failures or heave of bottoms of excavations; largely delays in construction, equipment loss, property damage

SOURCE: From Cedergren (1989), reprinted with permission from John Wiley & Sons, Inc.

4. Providing for the lubrication of slip surfaces

5. Trapping of groundwater in soil pores during earthquakes or other severe shocks, which leads to liquefaction failures

There are many different methods to mitigate the effects of groundwater on slopes. During construction of slopes, built-in drainage systems can be installed. For existing slopes, drainage devices such as trenches or galleries, relief wells, or horizontal drains can be installed. Another common slope stabilization method is the construction of a drainage buttress at the toe of a slope. In its simplest form, a drainage buttress can consist of cobbles or crushed rock placed at the toe of a slope. The objective of the drainage buttress is to be as heavy as possible to stabilize the toe of the slope and also have a high permeability so that seepage is not trapped in the underlying soil.

There can also be other indirect effects of groundwater on slopes. For example, when the groundwater evaporates at the toe of the slope, salt deposits can form. Both the high groundwater table and the surface salt deposits can kill or stunt the growth of plants and trees. Deposits of salts due to evaporation of groundwater are known as *evaporites*. In arid or semiarid regions where moisture is evaporating at the ground surface, they can form on or just beneath the ground surface. The three most common evaporites are gypsum, anhydrite, and sodium chloride (rock salt).

Examples of the effects of groundwater on slope vegetation are shown in Figs. 8.3 to 8.5. Note the relatively sparse vegetation near the toe of the slope. Most of the vegetation at the toe of the slope died due to the high groundwater table and salt deposits. As shown in Fig. 8.5, the salt deposits have formed a crust on top of the ground surface.

At the site shown in Figs. 8.3 to 8.5, there was formational material consisting of alternating beds of sandstone, siltstone, and claystone of the Eocene Santiago formation. Because of the adverse out-of-slope dip of the bedding, the slope was faced with fill (i.e., a stabilization fill) and a key was constructed at the toe of the slope. This condition is illustrated in Fig. 8.6, which presents a cross section through the slope. A subdrain was reportedly placed at the back of the fill key, although it may have been improperly installed or become clogged after construction. The arrows in Fig. 8.6 show the path of groundwater through the stabilization fill and up through the toe of the slope where it created the conditions shown in Figs. 8.3 to 8.5. The repair for this site consisted of the overexcavation of the entire slope face as shown in Fig. 8.7, the installation of a drainage system, and then the recompaction of the slope face with granular (free-draining) soil.

Figure 8.3   Toe of slope.

Figure 8.4   Toe of slope (arrow points to location of Fig. 8.5).

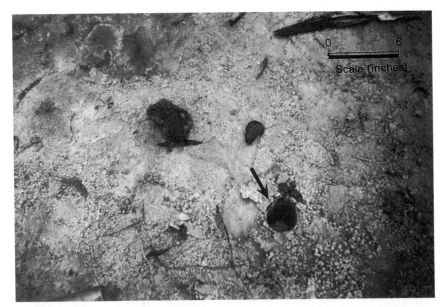

Figure 8.5 Close-up view of salt deposits.

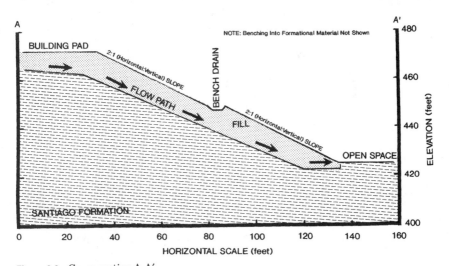

Figure 8.6 Cross section A-A'.

8.2 Moisture Migration through Floor Slabs

Moisture migration into buildings is one of the major problems faced by engineers, architects, and contractors. The problems associated with moisture migration into a structure are widespread and can be

Figure 8.7   Repair of site shown in Figs. 8.3 to 8.5.

expensive to fix. Frequently the inconvenience and cost of correcting the moisture problem results in litigation. In southern California, defects due to inadequate moisture barriers are a major cause of lawsuits against developers.

There are many different reasons for moisture migration into structures. Factors can include improper design, shoddy construction, or neglected maintenance. Moisture can migrate into the structure through the foundation, exterior walls, and through the roof. Four ways that moisture can penetrate a concrete floor slab are by water vapor, hydrostatic pressure, leakage, and capillary action (WFCA, 1984). Water vapor acts in accordance with the physical laws of gases, where water vapor will travel from one area to another area whenever there is a difference in vapor pressure between the two areas. *Hydrostatic pressure* is the buildup of water pressure beneath the floor slab; it can force large quantities of water through slab cracks or joints. *Leakage* refers to water traveling from a higher to a lower elevation due solely to the force of gravity, and such water can surround or flood the area below the slab. Capillary action is different from leakage in that water can travel from a lower to a higher elevation.

The controlling factor in the height of capillary rise in soils is the pore size (Holtz and Kovacs, 1981). Open graded gravel has large pore spaces and hence very low capillary rise. This is why open graded gravel is frequently placed below the floor slab to act as a capillary break (Butt, 1992).

Moisture that travels through the concrete floor slab can damage such floor coverings as carpet, hardwood, and vinyl. When a concrete floor slab has floor coverings, the moisture can collect at the top of the slab, where it weakens the floor-covering adhesive. Hardwood floors can be severely affected by moisture migration through slabs because they can warp or swell from the moisture.

Moisture that penetrates the floor slab can also cause musty odors or mildew growth in the space above the slab. Some people are allergic to mold and mildew spores. They can develop health problems from the continuous exposure to such allergens.

In most cases, the moisture that passes through the concrete slab contains dissolved salts. As the water evaporates at the slab surface, the salts form a white crystalline deposit, commonly called *efflorescence*. The salt can build up underneath the floor covering, where it attacks the adhesive as well as the flooring material itself. Figure 8.8 shows a photograph of salt deposits (white areas) and the growth of

Figure 8.8 Salt deposits and the growth of mold caused by moisture migration through a concrete slab (carpets have been removed).

Figure 8.9 Damage to wood flooring (arrow points to moisture stain; asterisk indicates upward warping of wood floor).

mold (dark areas) caused by severe moisture migration through a concrete slab-on-grade foundation.

Oliver (1988) believed that it is the shrinkage cracks in concrete that provide the major pathways for rising dampness. Sealing of slab cracks may be necessary in situations of rising dampness affecting sensitive floor coverings.

As an example of damage due to moisture migration through a concrete floor slab, Fig. 8.9 shows damage to wood flooring installed on top of a concrete slab. Only 6 months after completion of the house, the wood flooring developed surface moisture stains and warped upward in some places as much as 150 mm (0.5 ft). Most of the moisture stains developed at the joints where the wood planking had been spliced together. The joints would be the locations where most of the moisture penetrates the wood flooring. In Fig. 8.9, the arrow points to one of the moisture stains. The asterisk in Fig. 8.9 shows the location of the upward warping of the wood floor.

8.2.1   Moisture dome test and vapor flow rate

A valuable test to determine the quantity of moisture migrating through a concrete slab is the "moisture dome test." The moisture dome apparatus consists of a plastic cover and a preweighed dish of anhydrous calcium chloride. The test procedure is to put the dish containing calcium chloride on the slab surface, place the plastic cover over the calcium chloride, and then seal the plastic cover to the concrete slab. The moisture emitting from the concrete slab will be absorbed by the calcium chloride and the gain in weight of the calcium chloride can be used to calculate the vapor flow rate V, defined as

$$V = \frac{Q}{AT} \qquad (8.1)$$

where V = vapor flow rate or emission rate
$\quad Q$ = weight of water determined as the final minus initial weight of calcium chloride
$\quad A$ = surface area of the moisture dome test
$\quad T$ = time in days that the moisture dome test was performed

In practice, the vapor flow rate V is multiplied by 93 m^2 (1000 ft^2) to indicate that the vapor flow rate is for a 93 m^2 (1000 ft^2) slab area. The units for the vapor flow rate V are kg/day for a 93-m^2 area (lb/day for a 1000 ft^2 area). Rather than continuously repeat the 93 m^2 (1000 ft^2) slab area, the remainder of this book will express the vapor flow rates V as kg/day (lb/day).

The moisture dome test is a simple and inexpensive method to obtain the vapor flow rate at different slab locations and at different times of the year. Figure 8.10 shows the installation of the moisture

Figure 8.10 Moisture dome test.

dome test. Two possible reasons that the moisture dome test could underestimate the vapor flow rate are

1. The relative humidity could become greater inside than outside the apparatus.
2. The air inside the moisture dome is stagnant, but outside the moisture dome apparatus, circulating air could increase the removal of moisture from the slab surface.

8.2.2 Acceptable vapor flow rates

The *Moisture Test Kit Pamphlet* (Roofing Equipment Inc., undated) states that rubber, vinyl, or wood flooring can be safely installed if the vapor flow rate is less than 1.4 kg/day (3 lb/day). They further state that vinyl would probably be acceptable provided the vapor flow rate is less than 2.3 kg/day (5 lb/day).

Another source for acceptable vapor flow rates is the Carpet and Rug Institute. In their publication titled *Standard Industry Reference Guide for Installation of Residential Textile Floor Covering Materials,* the Carpet and Rug Institute (1995) states that:

> As a general guideline, an emission rate of 3 lbs [1.4 kg] or less is acceptable for most carpet. In the range of 3 to 5 lbs [1.4 to 2.3 kg], carpet with backings of porous construction can usually be installed successfully. An emission rate above 5 lbs is considered unacceptable.

For the example of the wood flooring previously described (see Fig. 8.9), the average vapor flow rate was 5.5 kg/day (12 lb/day). This vapor flow rate is 4 times greater than the maximum allowable value and was the reason for the moisture stains and warping upward of the wood floor (Fig. 8.9).

8.2.3 Structural design and construction details

For projects having moisture migration through floor slabs, the forensic engineer should investigate the moisture barrier and the gravel layer which are the main structural design and construction features used to prevent both water vapor and capillary rise through floor slabs. An example of below-slab recommendations (WFCA, 1984) are as follows:

> Over the subgrade, place 4 inches (10 cm) to 8 inches (20 cm) of washed and graded gravel. Place a leveling bed of 1 to 2 inches (2.5 to 5 cm) of sand over the gravel to prevent moisture barrier puncture. Place a moisture bar-

rier over the sand leveling bed and seal the joints to prevent moisture penetration. Place a 2 inch (5 cm) sand layer over the moisture barrier.

The gravel layer should consist of open graded gravel. This means that the gravel should not contain any fines and that all the soil particles are retained on the gravel size sieves. This will provide for large void spaces between the gravel particles. Large void spaces in the gravel will help prevent capillary rise of water through the gravel (Day, 1992d). In addition to a gravel layer, the installation of a moisture barrier (such as visqueen) will further reduce the moisture migration through concrete (Brewer, 1965).

8.2.4 Flat slab ceilings

In southern California, flat slab ceilings are commonly used as roof support for basements or below-grade garages. Different types of structures can be built on top of the flat slab. For example, the flat slab may support pedestrian walkways, planter areas, and lightweight structures. Figures 8.11 to 8.13 show the effects of moisture migration through the flat slab. In Fig. 8.11, the salt deposits were formed when moisture seeped down between the joints in the flat slab ceiling. Figures 8.12 and 8.13 show salt deposits caused by moisture migration through utility lines cut through the flat slab ceiling. As

Figure 8.11 Moisture migration through flat slab ceiling.

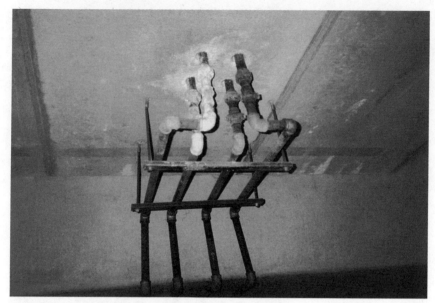

Figure 8.12 Moisture migration through flat slab ceiling.

Figure 8.13 Moisture migration through flat slab ceiling.

shown in Figs. 8.12 and 8.13, the salt deposits can cause corrosion of metal utility pipes. The key to minimizing moisture migration through the flat slab ceiling is to provide drainage of the flat slab surface, seal the joints, and prevent the cutting of holes through the flat slab ceiling.

8.3 Moisture Migration through Basement Walls

As in concrete floor slabs, water can penetrate basement walls by hydrostatic pressure, capillary action, and water vapor. If a groundwater table exists behind the basement wall, then the wall will be subjected to hydrostatic pressure, which can force large quantities of water through wall cracks or joints. A subdrain is usually placed behind the basement wall to prevent the buildup of hydrostatic pressure. Such drains will be more effective if the wall is backfilled with granular (permeable) soil rather than clay.

Another way for moisture to penetrate basement walls is by capillary action in the soil or the wall itself. By capillary action, water can travel from a lower to higher elevation in the soil or wall. Capillary rise in walls is related to the porosity of the wall and the fine cracks in both the masonry and, especially, the mortar. To prevent moisture migration through basement walls, an internal or surface waterproofing agent is used. Chemicals can be added to cement mixes to act as internal waterproofs. More common are the exterior applied waterproof membranes. Oliver (1988) lists and describes various types of surface-applied waterproof membranes.

A third way that moisture can penetrate through basement walls is by water vapor. As in concrete floor slabs, water vapor can penetrate a basement wall whenever there is a difference in vapor pressure between the two areas.

8.3.1 Damage due to moisture migration through basement walls

Moisture that travels through basement walls can damage wall coverings, such as wood paneling. Moisture traveling through the basement walls can also cause musty odors or mildew growth in the basement areas. If the wall should freeze, then the expansion of freezing water in any cracks or joints may cause deterioration of the wall. Similar to concrete floor slabs, the moisture that is passing through the basement wall will usually contain dissolved salts. The penetrating water may often contain salts originating from the ground or mineral salts naturally present in the wall materials. As the water evaporates at the interior wall surface, the salts form white crystalline deposits (efflorescence) on the basement walls. Figures 8.14 and 8.15 show photographs of the buildup of salt deposits on the interior surface of basement walls.

Figure 8.14 Two views of water migration through basement walls at a condominium complex in San Diego.

Figure 8.15 Two views of water migration through basement walls at a condominium complex in Los Angeles.

The salt crystals can accumulate in cracks or wall pores, where they can cause erosion, flaking, or ultimate deterioration of the contaminated surface. This is because the process of crystallization often involves swelling and considerable forces are generated. Another problem is penetrating water that contains dissolved sulfates, because they can accumulate and thus increase their concentration on the exposed wall surface, resulting in chemical deterioration of the concrete ("Aggressive Chemical Exposure," 1990).

The effects of dampness, freezing, and salt deposition are major contributors to the weathering and deterioration of basement walls. Some of the other common deficiencies that contribute to moisture migration through basement walls are as follows (Diaz, et al. 1994; Day, 1994d):

1. The wall is poorly constructed (for example, joints are not constructed to be watertight), or poor-quality concrete that is highly porous or shrinks excessively is used.

2. There is no waterproofing membrane or there is a lack of waterproofing on the basement-wall exterior.

3. There is improper installation, such as a lack of bond between the membrane and the wall, or deterioration with time of the waterproofing membrane.

4. There is no drain, the drain is clogged, or there is improper installation of the drain behind the basement wall. Clay, rather than granular backfill, is used.

5. There is settlement of the wall, which causes cracking or opening of joints in the basement wall.

6. There is no protection board over the waterproofing membrane. During compaction of the backfill, the waterproofing membrane is damaged.

7. The waterproofing membrane is compromised. This happens, for example, when a hole is drilled through the basement wall.

8. There is poor surface drainage, or downspouts empty adjacent the basement wall.

8.3.2 Structural design and construction details

The main structural design and construction details to prevent moisture migration through basement walls is a drainage system at the base of the wall to prevent the buildup of hydrostatic water pressure and a waterproofing system applied to the exterior wall surface.

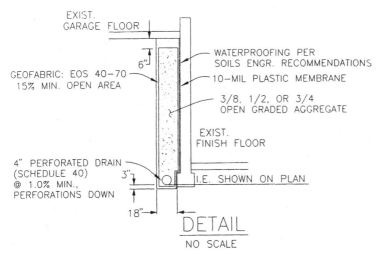

Figure 8.16 Typical waterproofing and drainage system for a basement wall. [*From Day (1996b), reprinted with permission from the American Society of Civil Engineers.*]

A typical drainage system for basement walls is shown in Fig. 8.16. A perforated drain is installed at the bottom of the basement wall footing. Open graded gravel, wrapped in a geofabric, is used to convey water down to the perforated drain. The drain outlet should be tied to a storm drain system.

The waterproofing system frequently consists of a high-strength membrane, a primer for wall preparation, a liquid membrane for difficult-to-reach areas, and a mastic to seal holes in the wall. The primer is used to prepare the concrete wall surface for the initial application of the membrane and to provide long-term adhesion of the membrane. A protection board is commonly placed on top of the waterproofing membrane to protect it from damage during compaction of the backfill soil. Self-adhering waterproofing systems have been developed to make the membrane easier and quicker to install.

8.3.3 Case study

This case study deals with the penetration of water through basement walls at a condominium project located in La Jolla, California. The author was retained by the homeowners association (HOA) to investigate and solve the moisture problems. The development consists of attached, two-story condominium units. Figure 8.17 shows a site plan of the building with the moisture problems. The basement walls, through which the water penetrated and flooded the living area, are located at the rear of the garage. The basement walls consist of concrete block supported by a reinforced concrete footing.

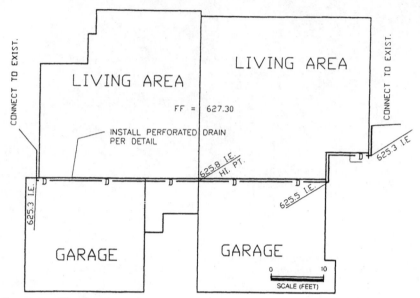

Figure 8.17 Site plan.

The building was constructed in 1975. During the 1980s and 1990s, it was reported that water periodically flooded the units during or immediately after extended periods of heavy rain. The cause of the problem was believed to be the temporary rise of the groundwater table. The main area of water intrusion seemed to be near the bottom of the basement wall, as observed by the location of the salt deposits.

During the excavation of the soil behind the basement walls, a subdrain was discovered. But the subdrain was not functioning because it was not connected to an underground drainage system, such as a storm drain line. The ends of the subdrain were simply buried in the ground. During the repair, this subdrain was completely removed. Also during the repair, there was a series of heavy rainstorms and water seeped into the excavation behind the basement wall. A sump pump was used to temporarily drain the water behind the basement wall during the repair.

Besides the defective subdrain, there were two other contributing factors in the flooding of the condominium units. One factor was that holes (for utility pipes) had been cut through the basement wall, as shown in Fig. 8.18. These holes allowed the groundwater to easily penetrate the wall. Another contributing factor was the lack of waterproofing. Figure 8.19 shows the installation of the waterproofing and new drainage system.

Figure 8.18 Holes in basement wall.

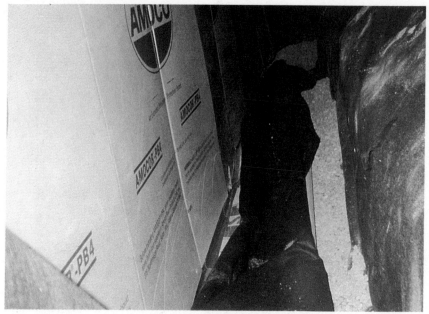

Figure 8.19   Repair.

In order to further investigate the source of the groundwater, samples of the groundwater were tested for fecal coliform. The presence of high concentrations of fecal coliform could indicate that a possible source of the water is from a leaking sewer or waste drain line. Test results indicated fecal coliform levels of less than 2 mpn per 100 mL, indicating that a leaking sewer line was not contributing to the rise in groundwater.

In summary, this case study deals with water penetration of basement walls. There were three main factors that caused the flooding of the units: (1) an inoperable subdrain, (2) leakage through utility holes that had been cut through the wall, and (3) a lack of waterproofing as shown in Fig. 8.18. The repair consisted of the installation of a new subdrain, sealing of holes cut in the basement wall, and the installation of a new waterproofing membrane (Fig. 8.19).

8.4   Pipe Breaks and Clogs

Pipes can be classified as either pressurized or nonpressurized. Common pressurized pipes are the water lines that provide potable water to the building occupants. An example of a nonpressurized pipe is a sewer line, which may be only intermittently filled with effluent.

Figure 8.20 Street collapse due to water line break.

A pressurized pipe break can introduce large volumes of water into the ground. This water can trigger collapse of loose soil, heave of expansive soil, or a raising of the groundwater table which may lead to slope instability. The large volume of water from a broken pressurized pipe can also erode and transport soil particles, leading to the development of voids below the structure. For example, Figs. 8.20 and 8.21 show two views of the collapse of a street due to the erosion of the base and subgrade caused by a pressurized pipe break.

Structural damage can also develop as a result of a nonpressurized break. At one building, there was substantial damage to the front bearing wall caused by the sudden subsidence of the ground surface. Subsurface exploration discovered that there was a sewer line that ran underneath the bearing wall. The top of the sewer line was broken and soil had slowly filtered down into the sewer line. Periodic cleaning of the sewer line probably helped to enlarge the void that was developing above the sewer line. Eventually the void collapsed, causing the ground surface subsidence and damage to the overlying bearing wall.

Besides pipe breaks, there can also be damage due to pipe clogs. There are many different ways that a pipe can become clogged. For example, the pipe can become clogged with debris or the overburden pressure can crush the pipe. Figure 8.22 shows an example of a

Figure 8.21   Close-up view of the street collapse.

Figure 8.22   Sewer main constructed through storm drain.

clogged storm drain. The storm drain was a critical drainage facility, and when it became clogged during a heavy rainstorm, there was extensive flooding and damage to the adjacent area. As shown in Fig. 8.22, the cause of the clog was a sewer main line that was constructed right through the center of the storm drain. The area below the sewer line became clogged with rocks, while the area above the sewer line was plugged by a plastic bottle (see Fig. 8.22). During the heavy rainstorm, the plastic bottle may have floated on top of the storm water and then become lodged in place above the sewer line. The plastic bottle shown in Figure 8.22 probably was the final plug for the storm drain which then led to the flooding and damage of the surrounding area.

In summary, the forensic engineer should determine if a pipe leak or clog contributed to the site damage. In some cases, building damage caused by a pipe leak may be covered by the owners' insurance policy. In other cases, the agency responsible for the maintenance of the pipe may be liable for the cost of repairs, such as in the case previously described where a sewer line was constructed through the center of a storm drain (Fig. 8.22).

Case study.   This case study describes the water main break that occurred at the Wesley Palms Retirement Center, which is a six-story building having appurtenant facilities. Reportedly, in the middle of the night on February 28, 1997, a 15-cm- (6-in.-) diameter main that provided water to the building broke. The water main was under high pressure and a tremendous volume of water flooded an electrical transformer room and then flowed through the entire first floor of the six-story building. Several million dollars of property damages were caused by the pipe break.

The location of the main line break was below an electrical transformer room. The asterisk in Fig. 8.23 indicates the top of the electrical transformer room. The excavation shown in Fig. 8.23 was used to remove and replace the broken water main.

An interesting feature of this water main break was the damage caused by high water pressure in the ground. Floor slabs and concrete flatwork were uplifted, even though in some cases they were reinforced and attached with dowels to the wall footings. In one case, an uplifted floor slab was located about 15 m (50 ft) from the pipe break. Since all the surrounding floor slabs were intact and undamaged, it was believed that the water from the pipe break migrated along permeable utility trenches. The water pressure in the permeable utility trench then increased to such a magnitude that it literally forced upward the floor slab.

Figure 8.23 Location of pipe break (asterisk indicates top of electrical transformer room).

Figure 8.24 shows a view of the damaged floor slab area inside the electrical transformer room. The upward movement of the floor slab damaged the electrical transformers to such an extent that they all had to be removed and replaced. Figure 8.25 shows the subsequent repair of the electrical transformer room floor slab. The repair consisted of the recompaction of the subgrade, placement of gravel, and then construction of a new concrete floor slab containing steel reinforcement.

8.5 Surface Drainage

The last section of Part 2 deals with surface drainage. Inadequate surface drainage can be an important factor in triggering soils problems. For example, as previously discussed, water ponding adjacent a foundation can contribute to expansive soil edge lift, the infiltration of ponding water can raise the groundwater table, and inadequate surface drainage can contribute to moisture intrusion problems through floor slabs or basement walls.

Figure 8.26 shows an example of poor surface drainage and ponding of water at a condominium complex. In Fig. 8.26, the condominium is visible along the right side of the photograph.

Figure 8.24   Damaged floor slab area, electrical transformer room.

Figure 8.25   Repair of electrical transformer slab.

Figure 8.26   Poor surface drainage at a condominium complex.

One method of assessing the adequacy of surface drainage is to perform a drainage survey. This consists of taking elevation points along the drainage paths and then calculating the drainage gradient. Figure 8.27 shows an example of a drainage survey performed by a land surveyor. The elevation at the entrance of the house was arbitrarily assumed to be elevation 100 ft, and then the elevations of the drainage paths around the house were determined. The squares in Fig. 8.27 indicate survey points, while the shaded circle near the front of the house indicates a drain. The drainage gradients were calculated as the change in elevation divided by the distance between the elevation points. In Fig. 8.27, the drainage directions are indicated by arrows with the drainage gradient (expressed as a percentage) written above the arrow. For the area around the house shown in Fig. 8.27, the drainage gradients varied from 0 percent (flat) up to 10.6 percent. The drainage survey shows that there is poor drainage along the right side of the house and that water ponds at the area of the wood fence.

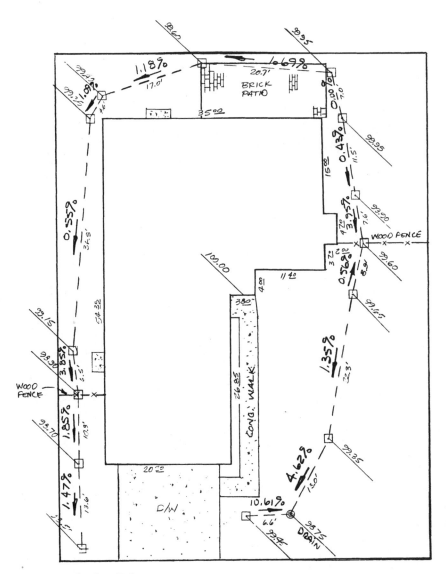

Figure 8.27 Surface drainage survey.

Part

3

Repairs

The photograph shows the removal of a damaged runway at the Reno/Tahoe International Airport and the construction of the base material for the new runway.

9

Repairs

The following notation is introduced in this chapter:

Symbol	Definition
D	Diameter of the pier (Sec. 9.5)
H	Distance from the slip surface to ground surface
k_a	Active earth pressure coefficient
L	Length (parallel to the slope) between rows of wood lagging and pipe piles (Sec. 9.4)
P	Force that each pipe pile resists
P_i	Required pier wall force that is inclined at an angle α
P_L	Lateral design force for each pier
S	Horizontal spacing between adjacent pipe piles (Sec. 9.4)
S	On-center spacing of the piers (Sec. 9.5)
u	Average pore water pressure along the slip surface
W	Total weight of the failure wedge
Z_T	Total length of the pier
Z_1	Depth to adequate bearing material

9.1 Development of Repair Recommendations

There are many forensic projects where the engineer is specifically hired to develop repair recommendations. For example, many owners had earthquake insurance and sustained damage to their buildings during the Northridge, California, Earthquake (magnitude 6.7, January 17, 1994). Once the claim is processed, the forensic engineer may be hired by the insurance company to determine the cause of damage (earthquake versus preexisting damage) and develop repair recommendations. In many cases, the forensic engineer will also be involved during the actual repair of the building through observation and testing.

Other projects where the forensic engineer may be hired to develop repair recommendations are for historic structures (Sec. 7.7). The forensic engineer could also be hired by individuals, corporations, financial institutions, real estate companies, homeowners associations (HOAs) or other types of owners in the private sector and given the assignment to fix the distressed or deteriorated structure. In these cases, the forensic engineer would first have to determine the cause of the problem and then prepare recommendations on the appropriate type of repair.

9.1.1 Projects involving civil litigation

For projects involving civil litigation, the forensic expert working for the plaintiff will usually have a different repair recommendation than the forensic expert working for the defendant. For example, the plaintiff's expert could have an extensive repair while the defendant's expert may provide a less rigorous or less expensive method to fix the problem. In many cases during the mediation process, the plaintiff's and defendant's forensic experts will agree on a compromise repair that is acceptable to the parties in the lawsuit. Once a repair has been agreed upon by the parties in the lawsuit, the settlement of the lawsuit is frequently a routine matter.

Problems naturally develop when the forensic expert working for the plaintiff has a significantly different and much more costly repair than the forensic expert working for the defendant. The cost of repair demanded by the plaintiff could even be based on the cost of completely demolishing and reconstructing the building, which was the basis for the $12 million damage award for the Trinity Church (ASCE, 1987; *Engineering News Record*, 1987), as discussed in Sec. 7.7.

The forensic engineer should realize that the recommended type of repair could impact his or her credibility at the time of trial. For example, at one project, there were serious deficiencies and damage to the foundation of a building. On one side, the footings for the foundation had not been constructed. At another location, a portion of the foundation had actually been constructed of soil, with a thin veneer of plaster used to hide the deficient condition. The defendant's expert stated in his deposition and at trial that the foundation was adequate and no repairs were required. The jury decided that this position was unreasonable and decided on a large monetary award to the plaintiffs. In addition, the jury decided that the costs of the plaintiff's investigation would be borne by the defendants. In this case, the unreasonable position of no repairs by the defendants impacted the jury's decision. A reasonable repair alternative by the defendant's expert would probably have resulted in a lower jury award, without the penalty of having to pay the plaintiff's investigation costs.

A second example deals with a large housing project having fill slope deformation which damaged top-of-slope walls and rear-yard appurtenances such as concrete patios and pool decks. The plaintiff's expert recommended that all of the fill slopes be strengthened with pier walls and the outer face of all slopes be removed and replaced with geogrid reinforced soil. The total cost of the repair was in excess of $100 million. Just prior to trial, the attorneys decided to have an unofficial "minitrial," with the merits of the case decided by a group of hired jurors. The consensus of the unofficial jurors was that the cost of repair was excessive because many of the fill slopes had not moved and in many areas, the damage was minor. In general, the unofficial jurors believed that the expensive repair was not necessary. Based on the outcome of this unofficial minitrial, the case settled for less than 10 percent of the requested amount. This case again illustrates that repairs considered unnecessary by the jury could impact the credibility of the forensic engineer.

The remainder of this chapter provides typical repair recommendations for the repair of damaged or deteriorated facilities. It is not possible to cover every type of geotechnical and foundation repair and the purpose of the remaining sections of this chapter is to provide examples of commonly used repair methods.

9.2   Repair of Slab-on-Grade Foundations

The most expensive and rigorous method of repair would be to entirely remove the slab-on-grade foundation and install a new foundation. This method of repair is usually reserved for cases involving a large magnitude of soil movement. Common types of new foundations are the reinforced mat or reinforced mat supported by piers (Coduto, 1994).

9.2.1   Reinforced mat

Figure 9.1 shows the manometer survey of a building containing two condominium units at a project called Timberlane in Scripps Ranch, California. The building shown in Fig. 9.1 was constructed in 1977 and was underlaid by poorly compacted fill that increased in depth toward the front of the building. In 1987, the amount of fill settlement was estimated to be 100 mm (4 in.) at the rear of the building and 200 mm (8 in.) at the front of the building. As shown in Fig. 9.1, the fill settlement caused 80 mm (3.2 in.) of differential settlement for the conventional slab-on-grade and 99 mm (3.9 in.) for the second floor. The rea-

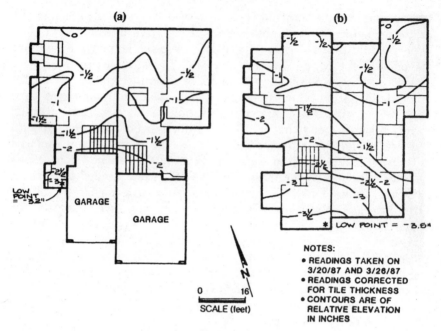

Figure 9.1 Manometer survey. (*a*) First floor; (*b*) second floor.

son the second floor had more differential settlement was because it extended out over the garage. Note in Fig. 9.1 that the foundation tilts downward from the rear to the front of the building, or in the direction of deepening fill. Typical damage consisted of cracks in the slab-on-grade, exterior stucco cracks, interior wallboard damage, ceiling cracks, and racked door frames. From Table 4.2, the damage was classified as severe. Because of ongoing fill settlement, the future (additional) differential settlement of the foundation was estimated to be 100 mm (4 in.).

In order to reduce the potential for future damage due to the anticipated fill settlement, it was decided to install a new foundation for the building. The type of new foundation for the building was a reinforced mat, 380 mm (15 in.) thick, and reinforced with no. 7 bars, 305 mm (12 in.) on center, each way, top and bottom.

In order to install the reinforced mat, the connections between the building and the existing slab-on-grade were severed and the entire building was raised about 2.4 m (8 ft). Figure 9.2 shows the building in its raised condition. Steel beams, passing through the entire building, were used to lift the building during the jacking process.

After the building was raised, the existing slab-on-grade foundation was demolished. The formwork for the construction of the reinforced mat is shown in Fig. 9.3. The mat was designed and constructed so that

Figure 9.2   Raised building.

Figure 9.3   Construction of mat foundation.

it sloped 50 mm (2 in.) upward from the back to the front of the building. It was anticipated that with future settlement, the front of the building would settle 100 mm (4 in.) such that the mat would eventually slope 50 mm (2 in.) downward from the back to the front of the building.

After placement and hardening of the new concrete for the mat, the building was lowered onto its new foundation. The building was then attached to the mat and the interior and exterior damages were repaired. Flexible utility connections were used to accommodate the difference in movement between the building and settling fill.

9.2.2   Reinforced mat with piers

A common foundation repair for structures subjected to settlement and/or slope movement is to remove the existing foundation and install a mat supported by piers. The mat transfers building loads to the piers, which are embedded in a firm bearing material. For a condition of soil settlement, the piers will usually be subjected to downdrag loads from the settling soil.

The piers are usually at least 0.6 m (2 ft) in diameter to enable downhole logging to confirm end bearing conditions. The piers can either be built within the building or be constructed outside the building with grade beams used to transfer loads to the piers. Given the height of a drill rig, it is usually difficult to drill within the building (unless it is raised). The advantages of constructing the piers outside the building are that the height restriction is no longer a concern and a large, powerful drill rig can be used to quickly and economically drill the holes for the piers.

Figure 9.4a shows a photograph of the conditions at an adjacent building at Timberlane. Given the very large magnitude of the estimated future differential settlement for this building, it was decided to remove the existing foundation and then construct a mat supported by 0.76-m- (2.5-ft-) diameter piers. The arrow in Fig. 9.4a points to one of the piers.

In order to construct the mat supported by piers, the building was raised and then the slab-on-grade was demolished. With the building in a raised condition, a drill rig was used to excavate the piers. The piers were drilled through the poorly compacted fill and into the underlying bedrock. The piers were belled at the bottom in order to develop additional end bearing resistance. After drilling and installation of the steel reinforcement consisting of eight no. 6 bars with no. 4 ties at 0.3-m (1-ft) spacing, the piers were filled with concrete to near ground surface. Figure 9.4b shows a close-up of the pier indicated in Fig. 9.4a. Note the bent steel reinforcement (no. 6 bars) at the top of the pier which is connected to the steel reinforcement in the mat.

At other projects, the piers have been constructed outside of the building. Figure 9.5 shows the construction of one such pier. Note the

Figure 9.4a Construction of mat supported by piers (the arrow points to one of the piers).

Figure 9.4b Close-up view of Fig. 9.4a.

Figure 9.5 Pier with grade beam at top.

steel reinforcement bending to one side at the top of the pier. This steel reinforcement was tied to the steel reinforcement in the grade beam, which then supported the new foundation.

Besides piers, the structure can also be underpinned with piles or screw or earth anchors (Brown, 1992). Greenfield and Shen (1992) present a list of the advantages and disadvantages of pier and pile installations.

9.2.3 Partial removal and/or strengthening of the foundation

A second type of repair for the slab-on-grade would be the partial removal of the damaged foundation and/or the strengthening of the foundation. This is usually a less expensive and rigorous method of repair than total removal and replacement of the foundation. The amount of soil movement is usually less for the case of partial removal and/or strengthening than for the case of total removal and replacement. The main objective of this type of repair is to fix the damaged foundation and then strengthen the foundation so that the damage does not reoccur. Partial removal and/or strengthening of the foundation is a common type of repair for damage caused by expansive soil (Chen, 1988).

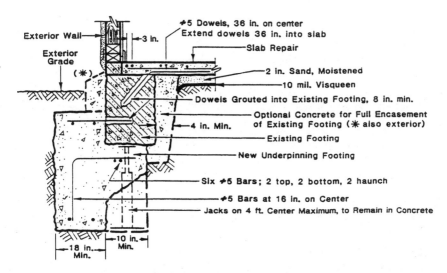

Exterior Wall
Exterior Grade
(✳)
←3 in.

✦5 Dowels, 36 in. on center
Extend dowels 36 in. into slab
Slab Repair
2 in. Sand, Moistened
10 mil. Visqueen
Dowels Grouted into Existing Footing, 8 in. min.
Optional Concrete for Full Encasement of Existing Footing (✳ also exterior)
Existing Footing
New Underpinning Footing
Six ✦5 Bars; 2 top, 2 bottom, 2 haunch
✦5 Bars at 16 in. on Center
Jacks on 4 ft. Center Maximum, to Remain in Concrete
4 in. Min.
10 in. Min.
18 in. Min.

Figure 9.6 Deepened perimeter footings. [*From Day (1996c), reprinted with permission from the American Society of Civil Engineers.*]

Figure 9.6 shows a cross section of a typical design for a deepened perimeter footing. The advantage of this type of repair is that the perimeter footing is strengthened and deepened. This can mitigate seasonal moisture changes and hence movement of the perimeter footings founded on expansive soil. The construction of the deepened perimeter footing starts with the excavation of slots in order to install the hydraulic jacks. After the hydraulic jacks have been installed (Fig. 9.6), the entire footing is exposed. Steel reinforcement is then tied to the existing foundation by dowels. The final step is to fill the excavation with concrete. The jacks are left in place during the placement of the concrete.

Figure 9.7 shows the installation of a deepened perimeter footing at a project called Oceana Mission, located in Oceanside, California. This building was subjected to landslide movement which cracked the perimeter footings in numerous places (Day, 1995b). Once the landslide movement had stopped, it was decided to strengthen the foundation by constructing a deepened perimeter footing. In Fig. 9.8, the steel reinforcement has been installed and the deepened footing is in the process of being filled with concrete.

For isolated interior concrete cracks, one method of repair is the strip replacement repair. Figure 9.9 shows a cross section of this type of repair. The construction of the strip replacement starts by cutting

Figure 9.7 Construction of deepened perimeter footing.

out by saw the area containing the concrete crack. Figure 9.9 recommends that a distance of 0.3 m (1 ft) on both sides of the concrete crack be saw-cut. This is to provide enough working space to install reinforcement and the dowels. After the new reinforcement (no. 3 bars) and dowels are installed, the area is filled with a new portion of concrete.

For those slabs that have unacceptable differential movement or are too badly damaged to be repaired by the strip replacement method, the entire slab can be removed and replaced. For example, Fig. 9.10 shows a photograph where the entire slab has been removed, except for the interior bearing wall footings. As shown in Fig. 9.10, the exposed expansive soil subgrade has been flooded to allow the clay to expand prior to construction of the new slab.

Figure 9.8 Construction of deepened perimeter footing.

9.2.4 Concrete crack repairs

The third type of repair is to patch the existing cracks in the concrete. Of the three general categories of repair, this is the least expensive repair method. This repair is usually recommended if the foundation has not excessively deformed (i.e., the foundation does not require releveling) and the slab-on-grade can accommodate the estimated future soil movement. The objective of this type of repair is to return the concrete slab to a satisfactory appearance and provide structural strength at the cracked areas. It has been stated (Transportation Research Board, 1977) that a patching material must meet the following requirements:

1. Be at least as durable as the surrounding concrete.

2. Require a minimum of site preparation.

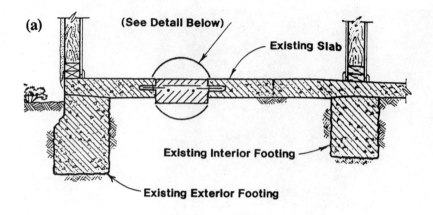

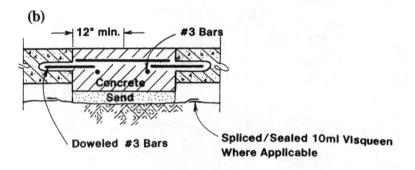

Saw–Cut 12" Each Side of Crack, Dowel #3 Bars 6"(min.) Into
Existing Slab. Provide 5" Concrete Section with #3 Bars, 12"
O.C. Both Ways . Underlay with 2" Moist Sand.   Where Visqueen
Exists,   Splice/Seal in a Replacement Section

Figure 9.9 Concrete crack repair. (*a*) Strip replacement of floor cracks; (*b*) strip replacement detail. [*From Day (1996c), reprinted with permission from the American Society of Civil Engineers.*]

3. Be tolerant of a wide range of temperature and moisture conditions.

4. Be noninjurious to the concrete through chemical incompatibility.

5. Preferably be similar in color and surface texture to the surrounding concrete.

Figure 9.11 shows a typical detail for concrete crack repair. If there is differential movement at the crack, then the concrete may require

Figure 9.10 Slab removal and replacement.

grinding or chipping to provide a smooth transition across the crack. The material commonly used to fill the concrete crack is epoxy. Epoxy compounds consist of a resin, a curing agent or hardener, and modifiers that make them suitable for specific uses. The typical range (3400 to 35,000 kPa, 500 to 5000 psi) in tensile strength of epoxy is similar to its range in compressive strength (Schutz, 1984). Performance specifications for epoxy have been developed (e.g., ASTM C 881-90 "Standard Specification for Epoxy-Resin-Base Bonding Systems for Concrete," ASTM, 1997a). In order for the epoxy to be effective, it is important that the crack faces be free of contaminants (such as dirt) that could prevent bonding. In many cases, the epoxy is injected under pressure so that it can penetrate the full depth of the concrete crack.

In summary, there are three general categories of repair for slab-on-grade: (1) removal and replacement of the foundation, (2) partial removal and/or strengthening of the foundation, and (3) concrete crack repairs. The type of repair depends on the magnitude of soil movement, the extent of damage, and the potential for future soil movement.

The most expensive and rigorous method of repair is the removal and replacement of the foundation. Common types of new foundations are the reinforced mat or reinforced mat supported by piers (Fig. 9.4).

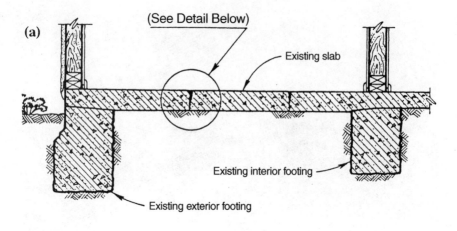

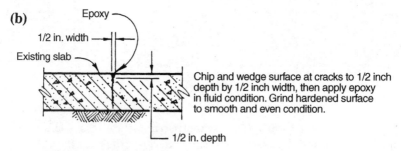

Figure 9.11   Concrete crack repair. (*a*) Epoxy repair of floor cracks; (*b*) detail of crack repair with epoxy. [*From Day (1996c), reprinted with permission from the American Society of Civil Engineers.*]

A second type of repair for the slab-on-grade is partial removal and/or strengthening of the foundation. The main objective of this type of repair is to fix the damaged foundation and then strengthen the foundation so that damage does not reoccur. Common methods used to accomplish this type of repair are deepened perimeter footings (Fig. 9.6) and/or removal and replacement of the slab (Fig. 9.10).

The third type of repair is to patch the existing cracks in the concrete. This repair is usually recommended if the foundation has not excessively deformed and if the slab-on-grade can accommodate the estimated future soil movement. The material commonly used to fill the concrete cracks is epoxy, which must meet performance and installation standards.

9.3   Other Foundation Repair Alternatives

Section 9.2 dealt with the strengthening or underpinning of the foundation in order to resist soil movement or bypass the problem soil. There are many other types of foundation repair or soil treatment alternatives (Brown, 1990, 1992; Greenfield and Shen, 1992; Lawton, 1996). In some cases, the magnitude of soil movement may be so large that the only alternative is to demolish the structure. For example, movement of the Portuguese Bend landslide in Palos Verdes, California, has destroyed about 160 homes. But a few homeowners refuse to abandon their homes as they slowly slide downslope. Some owners have installed steel beams underneath the foundation which are supported by hydraulic jacks that are periodically used to relevel the house. Other owners have tried bizarre stabilization methods, such as supporting the house on huge steel drums.

A more conventional repair alternative is to treat the problem soil. For example, fluid grout can be injected into the ground to fill in joints, fractures, or underground voids in order to stabilize settling structures (Graf, 1969; Mitchell, 1970). Another option is mudjacking, which has been defined as a process whereby a water and soil-cement or soil–lime cement grout is pumped beneath the slab, under pressure, to produce a lifting force which literally floats the slab to the desired position (Brown, 1992).

A commonly used in-place treatment alternative for foundation soils is compaction grouting, which consists of intruding a mass of very thick consistency grout into the soil, which both displaces and compacts the loose soil (Brown and Warner, 1973; Warner, 1982). Compaction grouting has proved successful in increasing the density of poorly compacted fill, alluvium, and compressible or collapsible soil. The advantages of compaction grouting are less expense and disturbance to the structure than foundation underpinning, and it can be used to relevel the structure. The disadvantages of compaction grouting are that it is difficult to analyze the results, it is usually ineffective near slopes or for near-surface soils because of the lack of confining pressure, and there is the danger of filling underground pipes with grout (Brown and Warner, 1973).

For expansive soil, mitigation options can include horizontal or vertical moisture barriers to reduce the cyclic wetting and drying around the perimeter of the structure (Nadjer and Werno, 1973; Snethen, 1979; Williams, 1965). Drainage improvements and the repair of leaky water lines are also performed in conjunction with the construction of the moisture barriers. Other expansive soil stabilization options include chemical injection (such as a lime slurry) into the soil below the structure. The goal of such mitigation measures is to induce a chemical mineralogical change of the clay particles which will reduce the soil's tendency to swell.

9.4 Repair of Surficial Slope Failures

Repair measures for surficial failures are often difficult to implement because of the problems of access and of working on the face of the slope. Commonly used repair measures are as follows:

9.4.1 Rebuild the failure area

An economical and relatively easy repair is to take the soil from the surficial slope failure and use it to rebuild the failed area. The organic matter, such as the trees and grass, is first separated from the surficial failure mass and disposed of off site. The design and repair consists of air drying the soil and then recompacting it in the failure area.

In most cases, especially for clays, this repair method is ineffective. One reason is that the clay will swell during the wet period and the benefits of the compaction will be lost. Another reason is because root reinforcement is destroyed during the surficial failure and the reestablishment of plants and trees may take several years.

9.4.2 Geogrid repair

Figure 9.12 shows the typical design using geogrid. The repair process is as follows:

1. *Removal of failure mass.* The first step is to remove and dispose of the surficial failure mass off site.

2. *Excavation of benches.* Benches are then cut into the hillside as shown in Fig. 9.12. The surficial failure may have created a slickened slide plane which can be eliminated by cutting benches into the underlying undisturbed ground. Benches provide favorable (i.e., not out-of-slope) frictional contact between the new fill mass and the horizontal portion of the bench. The benches are also used for the placement of the drainage system.

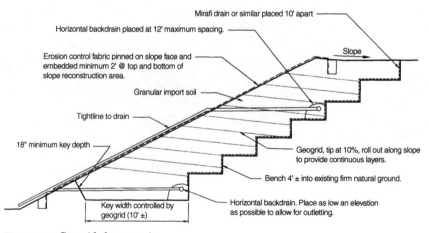

Figure 9.12 Geogrid slope repair.

Figure 9.13 Excavation of benches and installation of drains.

3. *Installation of drains.* After the benches have been excavated, drains are installed as shown in Fig. 9.12. The vertical drains are commonly placed at a 3-m (10-ft) spacing and are used to intercept seepage that may be migrating through the ground.

4. *Water disposal.* The horizontal drains collect the water from the vertical drains and dispose of the water off site. Figure 9.13 shows a photograph of the excavation of benches and installation of the drains.

5. *Rebuilding the slope.* The slope is rebuilt using layers of geogrid and compacted fill. The usual design specification is to compact the fill to a minimum of 90 percent of the Modified Proctor maximum dry density. It is common to import a granular soil to be used in the rebuilding process.

6. *Erosion control fabric.* At the completion of the rebuilding process, an erosion control fabric is pinned on the slope face and the slope face is replanted.

Figure 9.14 shows the repair of a surficial slope failure using geogrid. In the repair, geogrid acts as soil reinforcement, similar to the reinforcement effect of plant roots. Note in Fig. 9.12 that the geogrid is tipped back into the slope in order to get the geogrid as per-

Figure 9.14a Surficial slope repair with geogrid.

Figure 9.14b Completion of slope shown in Fig. 9.14a, with erosion control fabric on the slope surface.

pendicular to the potential failure surface as possible. The main design requirements are the type and vertical spacing of the geogrid. The design will depend on such factors as the shear strength of the import fill, the slope inclination, and the thickness of the potential failure mass. The design can be based on Eq. (6.1), where a geogrid force is included in the numerator.

9.4.3   Soil-cement repair

Figure 9.15 shows the soil-cement repair. The procedure for the soil-cement repair is the same as items 1 to 3 in the geogrid repair procedure. As shown in Fig. 9.15, instead of using geogrid, the granular import soil is usually mixed with 6 percent cement and compacted to at least 90 percent of the laboratory maximum dry density (Modified Proctor). After the soil-cement has been placed, planter areas are excavated within the slope face and the slope is landscaped.

For this repair, the cement increases the shear strength of the granular import soil which prevents future surficial failures. A major difficulty with this type of repair is the mixing of granular soil and cement. If the soil and cement are not thoroughly mixed, there can be uncemented zones that are susceptible to erosion or surficial failure.

In some cases, especially with flatter slopes, it may not be necessary to add cement to the import granular fill. In this case, the repair is identical to Fig. 9.15, except that the cement is not used.

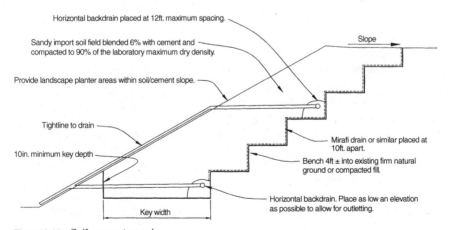

Figure 9.15 Soil-cement repair.

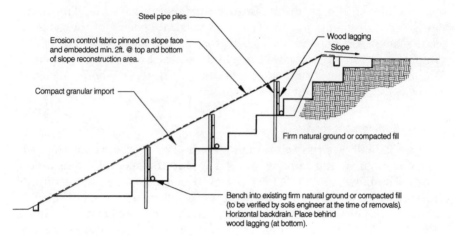

Figure 9.16 Pipe pile and wood lagging repair.

9.4.4   Pipe piles and wood lagging

Figure 9.16 shows a typical design using pipe piles and wood lagging. Pipe piles and wood lagging are probably the most frequently used repair method. As in the geogrid and soil-cement repairs, the slide debris is usually disposed of off site and benches are cut into natural ground.

Hollow galvanized steel pipe piles are either driven or placed in predrilled holes which are filled with concrete. Pressure-treated wood lagging is then placed behind the pipe piles and a drainage system is installed behind the wood lagging as shown in Fig. 9.16. The slope is then rebuilt with compacted fill. The slope face is provided with an erosion control fabric and landscaping.

A disadvantage of this repair is the usually low capacity (in terms of flexural strength) of the steel pipe piles. The wood lagging transfers the soil loads to the steel pipe piles. Because large soil forces can be generated in the surficial zone, the pipe piles frequently fail in bending. For example, Fig. 9.17 shows a photograph of a pipe pile and wood lagging system that failed. All that remains are the pipe piles that are bent downslope. The erosion control mat shown in Fig. 9.17 was placed on the slope face after the surficial failure.

The main reason that the pipe pile and wood lagging system frequently fails is because the system is not designed, but rather the size and spacing of the structural members are installed on the basis of the contractor's experience. In order to design the wood lagging system, it can be assumed that the wood lagging transfers all soil forces

Figure 9.17 Failure of wood lagging and pipe pile system (arrows point toward failed pipe piles).

to the steel pipe piles. If the force that each pipe pile resists is defined as P, Eq. (6.1) can be revised to include the resistance of the wood lagging and pipe piles as follows:

$$F = \frac{(P/S) + c'L + \gamma_b DL \cos^2 \alpha \tan \phi'}{\gamma_t DL \cos \alpha \sin \alpha} \qquad (9.1)$$

where S = horizontal spacing between adjacent pipe piles and L = the length (parallel to the slope face) between rows of wood lagging and pipe piles. Equation (9.1) can be rearranged to calculate the force P (which is inclined at the angle α) to be carried by each pipe pile:

$$P = F \gamma_t DSL \cos \alpha \sin \alpha - c'LS - \gamma_b DSL \cos^2 \alpha \tan \phi' \qquad (9.2)$$

Note that in Eqs. (9.1) and (9.2), the soil shear strength (c' = effective cohesion, ϕ' = effective friction angle) used in the analysis is the shear strength of the import granular fill. The distance D represents the estimated future zone of possible surficial instability, which could be assumed to be equal to the depth of the existing surficial failure. The factor of safety F applies to the soil parameters. Given an appropriate factor of safety in the design of the wood lagging and steel pipe piles, and a reliable determination of the import shear strength parameters, a low factor of safety (such as 1.2) could be used in the analysis. Also a low factor of safety could be justified because the analysis [Eqs. (9.1) and (9.2)] assumes seepage parallel to the slope face, which may be a conservative assumption given the installation of drains behind the wood lagging.

The maximum shear stress in the steel pipe piles is equal to P times cos α divided by the steel cross-sectional area of the pipe pile. The resultant force P can be assumed to act at a distance equal to $\frac{2}{3}D$ below ground surface. The maximum moment is calculated as P times cos α times the distance to the location of fixity of the pipe piles. If below the repair zone the pipe piles are imbedded in rock, the point of fixity will probably be close to the fill-rock contact. In many cases where the soil underlying the repair is soft, the point of fixity can be much deeper and hence the pipe piles will be subjected to a larger moment. This is the reason that the pipe pile system usually works best when there is rock or similar hard material just below the repaired area.

In addition to the design of the pipe piles, the wood lagging should be checked for shear and bending failure. A triangular soil pressure, increasing from zero at ground surface, can be assumed to act on the wood lagging. Using this pressure distribution, and knowing the total force P at each pipe pile, the maximum shear and maximum moment in the wood lagging can be calculated.

9.4.5   Suitability of repair method

Each surficial failure is different, and a single type of repair will not always be the most economical with an appropriate factor of safety. Rebuilding the failure area with the soil from the surficial failure will usually be the most economical treatment method. As previously mentioned, since the soil shear strength is not substantially changed, failure often recurs.

Geogrids can be used to increase the shear strength of the soil. However, there is the added cost associated with the installation of geogrids, such as purchasing and transporting the geogrids to the site. Using cement-treated soil is also labor-intensive because of the need to thoroughly mix the cement and the granular import soil.

Many contractors specialize in the pipe pile and wood lagging repair. As previously mentioned, in many cases this method of repair is not designed but rather the selection of members and spacing (S and L) are based on the contractor's experience. The economy of this method frequently occurs at the expense of an appropriate design, as shown by the failure in Fig. 9.17.

9.4.6 Summary

Four commonly used repair methods for surficial slope failures have been described. Besides causing damage to landscaping, surficial slope failures can destroy irrigation and drainage lines. A particularly dangerous condition can occur if the surficial slope failure mobilizes itself into a debris flow (Fig. 6.22).

The most economical repair method is to take the soil from the surficial slope failure and use it to rebuild the failed area. This repair method is usually ineffective because the soil shear strength is not substantially changed.

Reinforcing the soil can be accomplished by reconstructing the failure area by using geogrid (Fig. 9.12) or by adding cement to the import granular fill (Fig. 9.15). The repair includes the excavation of benches into the underlying undisturbed ground and drains are installed to intercept seepage that may be migrating through the ground.

Pipe piles and wood lagging (Fig. 9.16) are probably the most frequently used repair method. In many cases, this repair is not designed, but rather the installation is based on the contractor's experience. This frequently results in a new surficial failure because the pipe piles are overstressed and bend downslope (Fig. 9.17). Equation (9.2) can be used to calculate the force in the pipe piles, and this force can be used to check the shear and bending moment of the pipe piles.

9.5 Repair of Gross Slope Failures and Landslides

There are three basic approaches to the stabilization of a slope or landslide: (1) increase the resisting forces, (2) decrease the driving forces, or (3) rebuild the slope. Methods to increase the resisting forces include the construction of a buttress at the toe of the slope, or the installation of piles or reinforced concrete pier walls which provide added lateral resistance to the slope. Another technique is soil nailing which is a practical and proven system used to stabilize slopes by reinforcing the slope with relatively short, fully bonded inclusions such as steel bars (Bruce and Jewell, 1987).

Methods to decrease the driving forces include lowering the groundwater table by repairing leaky water pipes, improving surface drainage facilities, installing underground drains, or pumping groundwater from wells. Other methods to decrease the driving forces consist of removing soil from the head of the landslide or regrading the slope in order to decrease its height or slope inclination. The slope failure could also be rebuilt and strengthened by using geogrids or other soil reinforcement techniques (Rogers, 1992).

9.5.1 Pier walls

As previously mentioned, drilled piers are frequently used as a restraining system to stabilize slopes (Zaruba and Mencl, 1969). The terms *piers, shafts,* and *caissons* are sometimes used interchangeably by engineers. The common feature is that a cylindrical hole is drilled into the ground and then the hole is filled with concrete. The hole may be cased with a metal shell (a casing) in order to keep the hole from collapsing and sometimes to facilitate the cleaning of the bottom of the hole. The lower part of the hole may be belled out to develop a larger end-bearing area, thereby increasing the vertical load-carrying capacity of the drilled pier for a given allowable end-bearing pressure (Cernica, 1995b). General construction details for drilled piers are presented by Reese et al. (1981, 1985).

There are many different uses for drilled piers. Contiguous drilled piers are sometimes constructed to retain large open cuts more than 30 m (100 ft) deep (Abramson et al., 1996). When used to stabilize slopes, the drilled piers must pass through the slip surface and be embedded into a bearing stratum that can provide passive resistance to the destabilizing force transmitted by the unstable slope. Because soil arching can transfer lateral loads to the drilled piers, the piers are typically spaced a distance of 2 or 3 pier diameters apart.

In order to stabilize slopes, the required resistance of the pier wall may be very large. This can result in deep pier walls, having large pier diameters and substantial reinforcement to resist the overturning moment (Abramson et al., 1996). The construction cost can be reduced when the pier wall is combined with tiebacks. A tieback is normally constructed at the top of the pier by drilling a hole into the slope, installing a tieback into the hole, and then filling the anchoring portion of the hole with high-strength grout. The purpose of the tieback is to transfer a portion of the destabilizing force to a zone behind the slip surface. The tiebacks consist of high tensile-strength steel cables, tendons, or rods.

Two examples of the installation of pier walls to stabilize slopes are presented below:

Case study no. 1, Portuguese Bend landslide. A famous case of the failure of a reinforced concrete pier wall involves the Portuguese Bend landslide, located in the Palos Verdes peninsula, California. The landslide is about 105 ha (260 acres) in size with a typical thickness between 30 to 45 m (100 to 150 ft), (Watry and Ehlig, 1995). The Portuguese Bend landslide started to move in mid-1956 after the placement of a fill embankment for the extension of a roadway across the top of the landslide (Ehlig, 1992). The slip surface of the Portuguese Bend landslide occurs in bentonite layers within the Miocene Monterey formation. As previously mentioned, movement of the Portuguese Bend landslide has destroyed about 160 homes.

At the toe of the landslide, a cantilevered pier wall was constructed over a period of several months in 1957. A total of 23 reinforced concrete piers were installed. The concrete piers were 1.2 m (4 ft) in diameter and embedded 3 m (10 ft) below the slip surface. It was reported that the rate of movement of the landslide decreased 50 percent after installation of the concrete piers, although it has been argued that the slowing was due to other factors such as seasonal drying (Ehlig, 1986). Shortly after construction of the concrete pier wall, all of the piers failed by being plucked out of the ground, tilted downslope, or sheared off by the landslide movement (Watry and Ehlig, 1995).

Case study no. 2, fill slope failure. A second example of the construction of a pier wall to stabilize a slope involves a site located in San Diego, California. The site contains a single-family house with an 18-m- (60-ft-) high descending rear-yard fill slope inclined at approximately 1.5:1 (horizontal:vertical). In 1990, the homeowner discovered a pipe leak beneath the house. Water from the pipe leak infiltrated the fill mass and caused a slope failure. The slope failure resulted in 0.3 to 0.5 m (1 to 1.5 ft) of ground surface subsidence and the opening of tension ground cracks to approximately 100 mm (4 in.) in width. Figure 9.18 shows the ground cracks and foundation damage in the crawl space below the house.

The stabilization of the fill slope was achieved by the construction of a pier wall. Figures 9.19 and 9.20 show two views of the construction of the pier wall. Nineteen 0.9-m- (3-ft-) diameter piers, spaced 2.4 to 2.7 m (8 to 9 ft) on center, were installed through the fill mass and anchored in the underlying bedrock. Tiebacks were used to reduce the bending moments for the 10 most heavily loaded piers. The arrow in Fig. 9.19 points to the location of one of the inclined tieback anchors. The pier wall stabilized the fill slope by providing the needed lateral resistance. A retaining wall was included at the top of the piers. The final constructed condition is shown in Fig. 9.21.

Design of pier walls. One approach for the design of pier walls is to use slope stability analysis (Method A). The factor of safety of the slope, when stabilized by the pier wall, is first selected. Depending on

Figure 9.18   Ground cracks and foundation damage (arrows point to ground cracks).

Figure 9.19   Construction of pier wall (arrow points to tieback anchor).

Figure 9.20 Construction of pier wall.

Figure 9.21 Final constructed condition of the pier wall.

such factors as the size of the slope failure, the proximity of critical facilities, or the nature of the pier wall stabilization (temporary versus permanent), a factor of safety of 1.2 to 1.5 is routinely selected. The next step is to determine the lateral design force P_L that each pier must resist in order to increase the factor of safety of the slope up to the selected value.

Figure 9.22 shows an unstable slope having a planar slip surface inclined at an angle α to the horizontal. This mode of failure is similar to the Portuguese Bend landslide (Case Study 1) and the eventual wedge failure of the Desert View Drive fill embankment (case study in Sec. 6.5). The shear strength of the slip surface can be defined by the effective friction angle ϕ' and effective cohesion c'. The factor of safety F of the slope with the pier wall is derived by summing forces parallel to the slip surface:

$$F = \frac{\text{resisting forces}}{\text{driving force}} = \frac{c'L + (W\cos\alpha - uL)\tan\phi' + P_i}{W\sin\alpha} \quad (9.3)$$

where L = length of the slip surface
$\quad\;\; W$ = total weight of the failure wedge
$\quad\;\; u$ = average pore water pressure along the slip surface
$\quad\;\; P_i$ = required pier wall force that is inclined at an angle α as shown in Fig. 9.22.

The elements of Eq. (9.3) are determined as follows:

α and L The angle of inclination α and the length L of the slip surface are based on the geometry of the failure wedge.

W Samples of the failure wedge material can be used to obtain the wet density. From the wet density, the total weight W of the failure wedge (Fig. 9.22) can be calculated.

ϕ' and c' The shear strength parameters (ϕ' and c') can be determined from laboratory shear testing of slip surface specimens.

u By installing piezometers in the slope, the pore water pressure u can be measured.

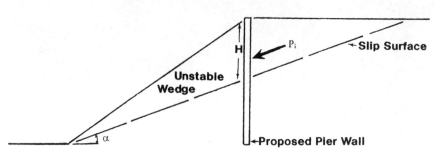

Figure 9.22 Design of pier wall for wedge slope failure.

F As previously mentioned, a factor of safety F of 1.2 to 1.5 is routinely selected for the slope stabilized with a pier wall.

The only unknown in Eq. (9.3) is P_i (inclined pier wall force, Fig. 9.22). The following equation can be used to calculate the lateral design force P_L for each pier having an on-center spacing of S:

$$P_L = SP_i \cos \alpha \qquad (9.4)$$

The location of the lateral design force P_L is ordinarily assumed to be at a distance of $\frac{1}{3} H$ above the slip surface, where H is defined in Fig. 9.22. The lateral design force P_L would be resisted by passive pressure exerted on that portion of the pier below the slip surface.

The above analysis to determine the lateral design force P_L for each pier was based on a wedge type of slope failure as illustrated in Fig. 9.22. If the slip surface is circular, then the method of slices can be used to calculate the lateral design force P_L. Circular slip surfaces are commonplace for slope failures in soil, such as Case Study 2.

A second approach for the design of pier walls is to use earth pressure theory (Method B) as illustrated in Fig. 9.23. The destabilizing pressure acting on the pier wall is assumed to be the active earth pressure. The active pressure is applied in a tributary fashion as if a continuous wall exists. The active pressure can be applied over the entire length Z_T of the pier.

The resistance to movement of the pier wall is from passive pressure. To account for arching, the passive pressure can be applied over 2 pier diameters (for pier spacing of $2D$ or greater). Note that the passive pres-

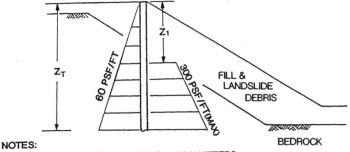

NOTES:
1) MAXIMUM PIER SPACING = 3 DIAMETERS

2) APPLY ACTIVE PRESSURE IN TRIBUTARY FASHION AS IF A WALL EXISTS. APPLY PASSIVE RESISTANCE OVER 2 PIER DIAMETERS

3) FOR TYPICAL CANTILEVER DESIGN, STRUCTURAL ENGINEER SHOULD ASSUME ABOUT 0.5 TO 1% ROTATION OF WALL IS REQUIRED TO MOBILIZE MAXIMUM PASSIVE RESISTANCE. TOTAL DEFLECTION SHOULD BE CALCULATED BY ADDING THIS ROTATION TO THE STRUCTURAL DEFLECTION FROM BENDING STRESS.

Figure 9.23 Design of pier walls using earth pressure theory.

sure begins at a depth of Z_1, which is defined as the depth to adequate lateral bearing material, determined from subsurface investigation.

To calculate the lateral design force P_L for each pier, the following equation is used:

$$P_L = 0.5 \, k_a \, \gamma_t Z_T^2 S \qquad (9.5)$$

where k_a = active earth pressure coefficient and γ_t = wet density. Because the active earth pressure is the actual driving pressure exerted on the pier wall, a factor of safety F can be included in Eq. (9.5). The factor of safety F could be the same value (1.2 to 1.5) as used in Method A (slope stability method).

As mentioned in Sec. 7.9, the value of k_a times γ_t is ordinarily referred to as the *equivalent fluid pressure*. The value of k_a can be calculated for a given shear strength of unstable slope material. In Fig. 9.23, the value of 60 psf/ft is equal to the equivalent fluid pressure recommended for the design of a pier wall to stabilize fill and landslide debris which moved during the Northridge earthquake (Day and Poland, 1996).

Earth pressure theory (Method B) can be used to calculate the lateral design force P_L when the slope deformation varies with depth (such as creep of clayey slopes), rather than failure on a distinct slip surface (Method A).

As previously mentioned, because soil arching can transfer lateral loads to the drilled piers, the piers are typically spaced a distance of 2 or 3 diameters apart. However, for some soils, this spacing may be too great. For example, Figs. 9.24 and 9.25 show pictures of soil movement around concrete piers. The soil shown in Figs. 9.24 and 9.25 was classified as a silty clay, having a liquid limit of 56 and a plasticity index of 32 (high plasticity). At this site, the clay was plastic enough to simply flow around the piers.

Table 9.1 presents recommendations for maximum spacing of piers versus soil or rock type. This table is based on the performance of existing pier walls. In general, the pier spacing should be decreased as the rock becomes more fractured or the soil becomes more plastic.

Construction of pier walls.   There are many references on the construction of piers (e.g., Reese et al., 1981, 1985). A major difficulty with the excavation of piers is access for drilling equipment. Because of the normal large size of the pier hole, great depth of excavation, and high resistance of the lateral bearing strata, a truck-mounted drill rig is required. This can make access a serious limitation for the construction of pier walls. If there is not enough room for the drill rig or if the topography is too steep, the option of using a pier wall for slope stabilization may have to be abandoned.

Figure 9.24 Soil movement around concrete piers.

Vertical loads can be imposed upon the pier by the tieback anchor or from the slope movement (Fig. 9.22). These vertical loads must be resisted by end-bearing and/or skin friction in the bearing strata. Loose soil or debris is frequently knocked into the hole during the construction of the pier. This generally occurs when the prefabricated steel cage is lowered into the pier hole. When the steel cage strikes the side wall of the pier, soil is inevitably knocked to the bottom of the pier hole. Once the steel cage is in place, it is nearly impossible to remove the accumulation of loose soil at the bottom of the pier. One solution to this problem is to design the piers to resist vertical loads only through skin friction between the concrete and the bearing strata. In this case, end-bearing resistance is neglected. Since lateral loads usually govern the depth of embedment, it may be economical to simply neglect end bearing and design the piers to resist vertical loads through skin friction.

Figure 9.25 Soil movement around concrete piers.

TABLE 9.1   Allowable Pier Spacing

Material type (1)	Largest on-center pier spacing (2)
Intact rock	No limit
Fractured rock	4D
Clean sand or gravel	3D
Clayey sand or silt	2D
Plastic clay	1.5D

NOTE: D = diameter of pier.

A small-diameter inclined hole is ordinarily used for the tieback anchor. This means that it will not be possible to visually observe the installation conditions of the tieback. Common problems are due to tiebacks not centered in the hole or inadequately grouted tieback zones which are unable to generate the required frictional resistance. Because of the numerous difficulties in construction of tieback anchors, field testing of the anchors is essential in order to be sure of an acceptable capacity.

Reducing Potential Liability

Image of Leaning Tower of Pisa in 1973 ASCE flyer promoting negotiation rather than competitive bidding for design services. (*Reprinted with permission from the American Society of Civil Engineers.*)

Chapter

10

Concluding Chapter

10.1 Future Outlook for Forensic Engineering

As infrastructure ages and marginal land is developed in urban areas, it is expected that there will be an increase in projects for forensic engineers. The trend in the United States is a broadening of civil liability, and it is anticipated that the future will bring more lawsuits against design professionals. Because of this trend, the final part of this book is devoted to the goal of reducing potential liability for design geotechnical and foundation engineers. Strategies for avoiding liability are also discussed.

In terms of the future for forensic engineers, Matson (1994) concludes:

> In many ways, the future looks bright. Brilliant minds are graduating from law schools in increasing numbers creating the potential for more litigation. The nature of litigation is more complex due to the reliance of society on technology. Governmental regulations in the areas of health, safety, and environment are more voluminous. All these trends mean that experts will be relied on with greater frequency. Indeed, expert witnessing is a growth industry.

10.2 Reducing Potential Liability

In their paper entitled "Claims Analysis from Risk-Retention Professional Liability Group," Janney et al. (1996) conclude that of all alleged errors or omissions, the highest percentage of claims was for design errors leading to delays or extras. Common reasons for claims due to delays or extra work on construction projects are as follows:

1. The client may request changes or additions.
2. There can be a misinterpretation of the statement of work, by the engineer, architect, contractor, or even the owner.
3. There can be omissions due to an improperly defined scope of work.
4. The requirements for the project may be poorly defined, overly optimistic, or unrealistic in terms of scheduling.
5. There may be a failure to take into account all of the risks of construction, especially with new or innovative techniques.
6. There may be outside influences (such as natural disasters) that affect the project.

But perhaps the most frequent cause of claims for delays or extra work is the bidding system (Gould, 1995). The goal of an owner or public agency is to obtain the lowest possible cost to construct the project. It is usually the low bidder who wins the contract. In order to prepare a low bid, contractors may not use a sufficient markup for overhead or profit. Once the job is started, contractors will hope to make a profit by claiming expenses due to delays or extra work. Owens (1993) states:

> All plans have errors and omissions. Contractors are experts at mining these hidden gold nuggets for extra profits. Public works-types projects seem especially vulnerable. Owners, whether airport authorities or wastewater treatment-plant officials, micro-manage. There is always some deadline that must be met. Contractors will stall, then, when reminded about the schedule, accelerate and claim extra costs.

The different objectives of the owner or public agency trying to construct a low-bid project and the contractor trying to make a profit through delays or extra work frequently result in litigation.

To reduce the number of claims due to design errors leading to delays or extras, Janney et al. (1996) have suggested the following:

1. Improved contracts between owners and constructors
2. Improved contracts between owners and design professionals and between design professionals
3. Partnering and dispute review boards (prompt resolution of emergency disputes)
4. Better control of realistic schedule for production of design documents
5. Better control of realistic turnaround time of all submittals

6. Quality control of design including regular "peer" checking, especially of details

7. Total quality management

10.3 Preparation of Reports and File Management

The purpose of the next two sections is to discuss the preparation of reports with the goal of reducing potential liability.

10.3.1 Report preparation

Wallace (1981) states that the purpose of a written report should be

1. To tell the client what we did for the money we are being paid.

2. To summarize the results of our work, providing a written record of the activities that have just concluded.

3. To satisfy the requirements of some agency or agencies ranging from local building departments to state or federal agencies.

4. When monitoring construction, to offer an engineering opinion, based upon our work, regarding the contractor's compliance with the applicable portions of the job plans and specifications.

And the purpose of a written report should not be

1. To guarantee, certify, or even accept the completed work. Guaranteeing is the business of contractors and manufacturers. Certifying is the business of banks and notaries. Accepting is the business of owners and architects.

2. To provide more information or opinions than contracted for between your firm and your client.

3. To rehash resolved disputes or arguments between the various parties involved in the project.

In general, a report should contain at least the following: (1) a statement of the type of project and the scope of work; (2) conclusions, recommendations, and calculations for significant design features; (3) description of the method of analysis; and (4) discussion of the design features and a clear description of any risk factors.

A common method of preparing a report is to use standardized formats, in which reports or letters are prepared by filling in the blanks. Wallace (1981) states this is a dangerous practice, because it tends to discourage the thought process. The danger is that the writer will make the conditions of the job fit the standardized form, instead of modifying the report or letter. There is also the tendency to forget those items not included on the standardized form. A better approach is to use a report outline and write the individual sections. Only a few sections should be standardized, such as the closure or limitation of liability clause.

10.3.2 Daily field reports

Daily field reports are common on construction projects. For example, daily field reports are used during the placement and compaction of fill. The report could be prepared by a field technician, engineer, or geologist. The daily field report would typically list the type of compaction equipment, the area of fill compaction, the results of field compaction tests, and observations during construction operations.

Daily field reports are generally recorded on duplicate forms, which are commonly titled "Report of Field Observations and Testing." The form usually has the company's name, address, and telephone number as well as spaces for the date, writer, project name and number, and client. The form usually has a main section to document field observations and testing. Some common items listed on this section of the daily field report are the contractor's operations, test results, and sketches or diagrams.

In many cases, the daily field report will be an in-house document that is used to prepare a construction progress report or final project completion report. Frequently a copy of the daily field report will be given to the contractor, supervisor, or client. Wallace (1981) states:

> The degree of care which goes into a daily field report should not vary depending upon who the reader will be, i.e., if the client gets a copy, do a good job; if it's only for the front office, do a sloppy job. It should be remembered that, in the event of a lawsuit, all information in all files related to the problem is available to all parties concerned. Your angrily scribbled blast at the grading contractor complete with swearwords—may eventually be read aloud to a judge and jury as well as a whole courtroom full of spectators.

The writer of a daily field report should prepare the document with the same care as a client's final report. Several rough drafts may be needed before the completion of the final report. A daily field report should not be a hasty summary of events, but a well-thought-out document that reflects field observations accurately.

10.3.3 File management

Examples of the contents of files include copies of reports sent to the client, daily field reports, correspondence, calculations, notes, and maps. As mentioned earlier, the information in the file related to the problem will be available to all parties during a lawsuit.

Some engineers believe that all documents, no matter how trivial or redundant, should be saved. Problems could occur if confidential, personal, or unrelated material is discovered in the file. At the other extreme are files devoid of all documents, except for reports issued to the client. This could cause problems because important photographs or testing of data, which may be vital to the defense, have been discarded. Neither extreme is desirable.

A file should be divided into separate compartments, such as confidential material, correspondence, testing, photographs, analysis, reports, and maps. Each individual compartment can be scrutinized for inappropriate material. Items that should be discarded include prior drafts of reports and letters, personal documents, rough notes or calculations (but keep the finished draft), and inappropriate or unrelated material. When a project is complete, the project manager should review the file and prepare it for the archives.

10.3.4 Examples of poor file management

Two examples of damaging documents discovered in geotechnical engineer's files are as follows:

In one defendant geotechnical engineer's file, a cartoon was discovered. The cartoon was titled "West Corner of Cribwall" and depicted a cross section through a house and rear-yard cribwall slope. A person was drawn looking out the rear window and saying, "I hope I don't lose my back yard!" The cartoon may demonstrate to the jury that this defendant was indifferent to the failing condition of the cribwall slope. Since juries may be unable to understand complex stability analyses or factors of safety, they may consider the cartoon as evidence of the engineer's negligence. The file should have been thoroughly reviewed on completion of the project and the cartoon discarded.

The second example concerns a large residential development with numerous fill slopes. These fill slopes experienced surficial instability as well as slope deformation, which required the demolition of one house. During the ensuing lawsuit the following daily field report, prepared by the geotechnical engineer during the construction of the site, was discovered in the file:

> [The owner] was still concerned about the fact that the material in the slope was not what was indicated in the plans, and I finally assured him that we would write a letter with our opinion that the material was adequate, and had adequate strength, such that deep-seated slope failure was not likely. This seemed to satisfy him at this time. [The city inspector] still appeared to be very concerned about the material, and could not be con-

vinced that there was not something drastically wrong done during construction of the slope.

The memo is a candid admission of the fact that the slope was not built in accordance with the approved plans. Such self-incriminating memos could increase potential liability. Perhaps a better approach would have been to accurately document field conditions by describing the reasons, limits, and remedial measures for the area not being constructed in accordance with the plans.

10.4   Engineering Jargon, Superlatives, and Technical Words

Geotechnical and foundation engineers, like most professionals, have words or phrases that have special meaning. This engineering jargon can cause problems. For example, some words have legal implications, others may be misunderstood by the layman. This can result in increased liability on the part of the geotechnical engineer.

Attorneys have said that a report speaks for itself. Years after a project is complete, the official record will be the written report. The words in the report can be scrutinized and interpreted. It is essential that the report be clear and concise. Careful selection of words and phrases can be important in avoiding ambiguity. For example, two words, *inspect* and *observe,* may mean exactly the same to an author, but could have significantly different meaning in a court of law. The purpose of the following section is to help the engineer avoid those words that can increase legal liability.

10.4.1   Engineering jargon

Certification, guarantee, ensure, or warrant.   Certification reports are common in engineering practice. One example is a certification report for the compaction of fill. A typical report title and text are as follows:

> *Title:* Report of Certification of Compacted Fill
> *Text:* This report was prepared to certify the degree of compaction of fill materials placed and compacted at the site, and in conformance with the attached grading specifications.

As the fill at the development is being certified, the geotechnical engineer will most likely be sued if damage occurs due to fill problems. This is because the words *certification, guarantee, ensure,* and *warrant* can have legal definitions that increase the liability of the geotechnical engineer. For example, Narver (1993) states:

> In many states, a certification is synonymous with a warranty. If you certify that a project was constructed in accordance with the approved plans

and specifications, you could be guaranteeing contractor compliance with all specifications. If it is later shown that the contractor was less than 100% in compliance, you might be held liable for the consequences on the theory that other parties may have relied upon your certification, or that you warranted another party's work.

Instead of certifying the fill, the report title and text could be as follows:

Title: Report of Observations of Compacted Fill
Text: This report presents the results of fill compaction tests and observations during the placement and compaction of fill at the site.

Because of the potential increased liability associated with *certification, guarantee, ensure,* and *warrant,* geotechnical engineers should avoid the use of these words.

Controlling, inspecting, supervising. A primary responsibility of engineers is to review the work of junior engineers, builders, and contractors. In fact, registration as a civil engineer cannot be obtained unless the engineer-in-training has responsible charge. This may cause engineers-in-training to use words such as *inspect* or *supervise* in reports to comply with the responsible charge requirement. But these words have legal meaning. For example, Narver (1993) states:

Observe means "to notice with care; to be on the watch respecting." By contrast, inspection has been determined by several courts to have a much broader meaning. A Missouri court found the word inspection to mean "to examine carefully or critically, investigate and test officially." A Pennsylvania court went a step further, finding that there was no substantial difference between the services of "inspection" and "supervision." This last definition is significant because it places upon a design professional who inspects a project the responsibilities of a supervisor. Consequently, by agreeing to perform "inspection services" or using the term "inspection" in conjunction with construction observation, the design professionals can potentially subject themselves to additional claims from the owner, contractors, workers, and the government for problems relating to compliance with plans and site safety.

Geotechnical and foundation engineers should avoid the use of *control, inspect,* and *supervise,* unless they carefully or critically investigate each step of the work. Alternatives include words such as *observe, review, study, look over* (the job), and *perform tests.*

10.4.2 Superlatives

Warriner (1957) states that *comparison* is the name given to the change in the form of adjectives and adverbs when they are used to

TABLE 10.1   Degrees of Comparison

Positive (1)	Comparative (2)	Superlative (3)
Good	Better	Best
Long	Longer	Longest
Many, much	More	Most

compare the qualities of words they modify. There are three degrees of comparison: positive, comparative, and superlative. Table 10.1 lists several examples.

Commonly used superlatives are *must, always* (or *all*), *never,* and *none.* Engineers use superlatives because of their education and training. They are taught to be exact, with the result that superlatives are routinely used in everyday correspondence.

An example from the writer's experience involved the construction of a house on expansive soil. The geotechnical engineering firm involved with the construction stated in their final compaction report:

> In addition to capping lots as described above, the cut portion of cut/fill transition lots was undercut at least 3 feet and replaced with compacted fill so that all lots are either entirely cut or entirely fill lots.

In terms of expansive soil, the soil engineering firm stated that all lots have very low to low expansion potential. In both statements, the geotechnical engineering firm used superlatives, such as "all," "at least," and "entirely."

Shortly after construction, the house began to experience cracking. The same geotechnical engineering firm investigated the damage to the house. They prepared a report on the damage to the house and stated the following:

> In our opinion, the distress is likely related to a combination of minor soil settlement within the less densely compacted fill soils, the presence of highly expansive soils near grade within the remaining portion of the building pad and the differing soil settlement characteristics of the fill soils and formation soils due to the cut portion of the lot not being undercut.

The geotechnical engineering firm was candid. They admitted that the cut portion of the lot was not entirely undercut and there was highly expansive soil near grade in the building pad. Superlatives in the final compaction report had an impact on the legal liability of the geotechnical engineering firm. Instead of using superlatives, the geotechnical engineering firm could have stated that, on the basis of observations during grading, the lots were graded as either cut or fill lots.

10.4.3 Technical words

The geotechnical and foundation engineering professions have numerous technical words that have limited or unusual meanings. Sweet (1970) states that in a court of law, technical words will be interpreted as usually understood by persons in the profession or business to which they relate, unless clearly used in a different sense. But there may not be a consensus of opinion as to the exact meaning of a technical word. Words such as *critically expansive* and *collapsible soil* can be defined by specific parameters, but a jury may focus on the words *critically* and *collapsible* to form their conclusions. To a jury, if something sounds bad (*critically* or *collapsible*), then they think it most likely is bad. Such technical words should be clearly defined or avoided so that false impressions are not formed.

10.5 Strategies to Avoid Civil Liability

Many engineers believe that if they practice perfect engineering or conform to the standard of care, they will be immune to civil liability. Unfortunately, if a project develops problems, all engineers may be named in the lawsuit, regardless of their innocence. In the United States, there are attorneys who sue everyone involved with a damaged project to secure as much money for their clients as possible (Meehan et al., 1993).

Today, a practicing engineer cannot perform services without the fear of a lawsuit. A lawsuit that could result in substantial financial loss is every engineer's nightmare. Even if the engineer is vindicated, the drain on resources and the stress of deposition and trial can be debilitating (Day, 1992e).

One well-documented case history involves L'Ambiance Plaza, which collapsed during construction in Bridgeport, Conn., killing 28 workers. Fairweather (1992) dealt with the personal effect of the collapse on James O'Kon of O'Kon and Co., who was the engineer of record for L'Ambiance Plaza. James O'Kon summarized his experience:

> No suits or claims were filed against my firm. The contractors did not survive, the developer did not survive. O'Kon and Co. survived. I was declared guilty by the press, investigated for murder, almost lost my professional license, and threatened with business failure by a federal judge.

There are ways to limit one's potential liability. The following discussion is primarily focused toward practicing geotechnical and foundation engineers who provide consulting services in small to midsize companies.

10.5.1 Assessing risk

Engineers who have practiced in a locality for a long time can develop consulting services that have a low risk. For example, an engineer may provide consulting service to the local or county government, which may have set procedures of trying to negotiate a fix for problems, rather than sue. Other examples of low-risk activities could be consulting work for longtime clients, where a personal relationship has been developed. Such a relationship results in solving problems, instead of trying to fix blame in a court of law. Often the best approach to avoiding a lawsuit is to work with the client to solve the problems as they develop.

Byer (1992) defines clients as low-budget speculators, high-budget speculators, remodelers, mass builders, and the professional builder. He assesses the risk of working for each of these clients and concludes that the type of client in a project can affect your liability on the project. He suggests that the speculators are high-risk clients because they will tend to cut corners, while the professional builder generally has a low risk of failure or problems because of the insistence on quality work.

Besides the type of client, the type of project can also affect liability. Some types of projects with a very high potential for liability are as follows:

1. *Toxic waste.* The failure to find and treat all of the toxic waste is always a possibility.

2. *Condominium complexes.* Performing geotechnical or foundation design engineering services for these developments is very risky, especially in California, because if damage develops, the developer could be held strictly liable for that damage, regardless of negligence. When a developer is sued, cross-complaints against the design engineers are inevitable.

3. *Trench excavations.* There are probably more injuries and deaths due to trench excavations than any other construction activity. Frequently, trenches are excavated for long distances and encounter many different soil and groundwater conditions, making the design or observation of shoring difficult.

10.5.2 Insurance

One way to deal with potential liability is to purchase errors-and-omissions insurance. If sued, the case can be turned over to the insurance company. After the deductible is exhausted, the insurance company will have to pay the attorney's fees, provide the legal defense, and pay judgments up to the limits of the policy.

Insurance, however, can be very expensive, especially for high-risk activities such as geotechnical and foundation engineering. There is also the problem that if an individual or company is sued, the rates could significantly escalate. Some engineers believe that having insurance invites lawsuits. For example, a developer actually stated that he was more interested in the adequacy of the engineering consultant's insurance policy rather than professional abilities. He was quoted as saying, "It's your [the engineer's] insurance policy I want" (Meehan et al., 1993).

For uninsured defendants, Patton (1992) has stated:

> Since the financial condition and insurance policy limits are factors to be considered in determining whether or not a proposed settlement is made in "good faith," consideration should be made in keeping the operating corporation as empty as possible and without insurance coverage. Even where the plaintiff's damages are accepted as great and the liability certain, a disproportionately low settlement figure is often reasonable in the case of a relatively insolvent and uninsured defendant. This requires strong wills, steady nerves and determination on the part of the uninsured defendants.

10.5.3   Limitation of liability clauses

Another common strategy for avoiding lawsuits is limitation of liability clauses. These are usually included in the fine print of the contract between the consulting engineer and the client. An example of a limitation of liability clause is as follows:

> *Limitation of liability:* This contract contains a limitation of liability clause as follows: In the event of a legal proceeding or other action arising from or related to this agreement or the consultant's performance hereunder, the consultant's total liability to the client, whatever form the action, shall be limited to 1.4 times the professional fees collected by the consultant for the project.

Some of the problems with limitation of liability clauses are

1. The limitation of liability clause may not be judgment-proof. Case law continually evolves, and it is possible that a limitation of liability clause will be ruled invalid by a judge or court of appeals.
2. The client may be simply unwilling to sign the contract with a limitation of liability clause. Then the choice is whether to accept the increased risk of the job or refuse the assignment.
3. There is always the possibility of a lawsuit by a third party, such as an injured bystander, who is not a party to the contract.

In summary, a practicing geotechnical and foundation engineer cannot perform services without the fear of a lawsuit. Some strategies for avoiding civil liability include assessing the risk before accepting the assignment, purchasing errors-and-omissions insurance, and using limitation of liability clauses in contracts.

Recommended Practices for Design Professionals Engaged as Experts in the Resolution of Construction Industry Disputes*

Preamble

Experts are vitally important to contemporary American jurisprudence. They review and evaluate complex technical issues and explain their findings and opinions to lay triers of fact for the latter's consideration in reaching a verdict.

Experts retained by opposing parties may disagree. In all instances, such disagreements should emanate only from differences in professional judgment.

These recommendations have been developed from the belief that adherence to them will help experts provide to triers of fact substantiated professional opinions unbiased by the adversarial nature of most dispute resolution proceedings. The organizations which endorse these recommendations do not require any individual to follow them.

Recommendations

It is the obligation of an expert to perform in a professional manner and serve without bias. Toward these ends:

---

*Reprinted with permission from ASFE: The Association of Engineering Firms Practicing in the Geosciences.

1. The expert should avoid conflicts of interest and the appearance of conflicts of interest.

COMMENTARY: Regardless of the expert's objectivity, the expert's opinion may be discounted if it is found that the expert has or had a relationship with another party which consciously or even subconsciously could have biased the expert's services or opinions. To avoid this situation, experts should identify the organizations and individuals involved in the matter at issue, and determine if they or any of their associates have or ever had a relationship with any of the organizations or individuals involved. Experts should reveal any such relationships to their clients and/or clients' attorneys to permit them to determine whether or not the relationships could be construed as creating or giving the appearance of creating conflicts of interest.

2. The expert should undertake an engagement only when qualified to do so, and should rely upon other qualified parties for assistance in matters which are beyond the expert's area of expertise.

COMMENTARY: Experts should know their limitations and should report their need for qualified assistance when the matters at issue call for expertise or experience they do not possess. In such instances, it is appropriate for experts to identify others who possess the required expertise, and to work with them. Should an expert be asked to exceed his or her limitations and thereafter be denied access to other professionals, and should the expert be requested to continue association with the case, the expert should establish which matters he or she will and will not pursue; failing that, the expert should terminate the engagement.

3. The expert should consider other practitioners' opinions relative to the principles associated with the matter at issue.

COMMENTARY: In forming their opinions, experts should consider relevant literature in the field and opinions of other professionals when such are available. Experts who disagree with the opinion of other professionals should be prepared to explain to the trier of fact the differences which exist and why a particular opinion should prevail.

4. The expert should obtain available information relative to the events in question in order to minimize reliance on assumptions, and should be prepared to explain any assumptions to the trier of fact.

COMMENTARY: The expert should review those documents, such as tenders and agreements, which identify the services in question and any restrictions or limitations which may have applied. Other significant information may include codes, standards and regulations affecting the matters in dispute, and information obtained through discovery procedures. If pertinent to the assignment, the expert should also visit the site of the event involved and consider information obtained from witnesses. Whenever an expert relies on assumptions, each assumption should be identified and evaluated. When an assumption is selected to the exclusion of others, the expert should be able to explain the basis for the selection.

5. The expert should evaluate reasonable explanations of causes and effects.

COMMENTARY: As necessary, experts should study and evaluate different explanations of causes and effects. Experts should not limit their inquiry for the purpose of proving the contentions advanced by those who have retained them.

6. The expert should strive to assure the integrity of tests and investigations conducted as part of the expert's services.

COMMENTARY: Experts should conduct tests and investigations personally, or should direct their performance through qualified individuals who should be capable of serving as expert or factual witnesses with regard to the work they performed.

7. The expert should testify about professional standards of care only with knowledge of those standards which prevailed at the time in question, based upon reasonable inquiry.

COMMENTARY: When a design professional is accused of negligence, the trier of fact must determine whether or not the professional breached the applicable standard of care. A determination of the standard of care prevailing at the time in question may be made through investigation, such as the review of reports, records, or opinions of other professionals performing the same or similar services at the time in question. Expert witnesses should identify standards of care independent of their own preferences, and should not apply present standards to past events.

8. The expert should use only those illustrative devices or presentations which simplify or clarify an issue.

COMMENTARY: The attorney who will call the expert as a witness will want to review and approve illustrative devices or presentations before they are offered during testimony. All illustrative devices or presentations developed by or for an expert should demonstrate relevant principles without bias.

9. The expert should maintain custody and control over whatever materials are entrusted to the expert's care.

COMMENTARY: The preservation of evidence and the documentation of its custody and care may be necessary for its admissibility in dispute resolution proceedings. Appropriate precautions may in some cases include provision of environmentally controlled storage.

10. The expert should respect confidentiality about an assignment.

COMMENTARY: All matters discussed by and between experts, their clients and/or clients' attorneys should be regarded as privileged and confidential. The contents of such discussions should not be disclosed voluntarily by an expert to any other party, except with the consent of the party who retained the expert.

11. The expert should refuse or terminate involvement in an engagement when fee is used in an attempt to compromise the expert's judgment.

COMMENTARY: Experts are employed to clarify technical issues with objectivity and integrity. Experts should either refuse or terminate service when they know or have reason to believe they will be rewarded for compromising their objectivity or integrity.

12. The expert should refuse or terminate involvement in an engagement when the expert is not permitted to perform the investigation which the expert believes is necessary to render an opinion with a reasonable degree of certainty.

COMMENTARY: It is the responsibility of experts to inform their clients and/or their clients' attorneys about the scope and nature of the investigation required to reach opinions with a reasonable degree of certainty, and the effect which any time, budgetary or other limitations may have. Experts should not accept or continue an engagement if limitations will prevent them from testifying with a reasonable degree of certainty.

13. The expert should strive to maintain a professional demeanor and be dispassionate at all times.

COMMENTARY: Particularly when rendering testimony or during cross-examination, expert witnesses should refrain from conducting themselves as though their service is a contest between themselves and some other party.

Endorsing Organizations (as of July 1, 1993)

ASFE/Professional Firms Practicing in the Geosciences, February 15, 1988

American Academy of Environmental Engineers, March 15, 1988

American Association of Cost Engineers, July 9, 1988

American Association of Engineering Societies, December 7, 1989

American Congress on Surveying and Mapping, January 24, 1989

American Consulting Engineering Council, January 18, 1988

American Council of Independent Laboratories, April 8, 1988

The American Institute of Architects, March 14, 1988

American Institute of Certified Planners, April 25, 1987

American Institute of Professional Geologists, January 25, 1992

American Nuclear Society, November 2, 1988

American Public Works Association, April 4, 1989

American Society of Agricultural Engineers, October 20, 1988

American Society of Certified Engineering Technicians, January 21, 1989

American Society of Civil Engineers, October 23, 1988

American Society of Consulting Planners, April 30, 1988

American Society of Landscape Architects, August 15, 1987

American Society of Mechanical Engineers, March 15, 1990

American Society of Safety Engineers, June 1, 1988

American Tort Reform Association, September 26, 1989

Association of Energy Engineers, November 4, 1988

Association of Engineering Geologists, April 23, 1988

California Geotechnical Engineers Association, June 23, 1988

Illuminating Engineering Society of North America, June 7, 1988

International Federation of Consulting Engineers, February 10, 1990

Interprofessional Council on Environmental Design, March 17, 1988

National Academy of Forensic Engineers, January 26, 1988

National Society of Professional Engineers, January 21, 1988

Structural Engineers Association of Illinois, December 6, 1988

Washington Area Council of Engineering Laboratories, June 23, 1988

B

Case Management Order

The following is a case management order, issued by a superior court judge (public document). It provides an example of the judge's orders and deadlines for a project involved in civil litigation.

Special thanks to Billie Jaroszek, attorney at law, from the law firm of Flick, Jaroszek, Roth, and Kennedy, who provided the original copy of the Case Management Order.

1

2

3

4

5

6

7

8   SUPERIOR COURT OF THE STATE OF CALIFORNIA

9   FOR THE COUNTY OF SAN DIEGO

10   JOHN DOE, et al.   )   Case No. 00000000
)
11   Plaintiffs,   )   [PROPOSED] CASE MANAGEMENT
)   ORDER
12   v.   )
)
13   JANE DOE, et al.   .   )
)
14   Defendants.   )
      _____)
15   )
      AND RELATED CROSS-ACTIONS   )
16   _____)

17

18   Having considered the matter, reviewed the pleadings and heard oral argument of counsel for

19   the purposes of managing the above-captioned action and for good cause shown:

20   IT IS HEREBY ORDERED that the provisions set forth herein shall be the case management

21   order for the above-captioned action and all actions later consolidated herewith.

22   1.   **NEW PARTIES.** The cut-off date for adding any new party is October 1, 1997, and

23   no new party may be added thereafter without leave of court.

24   2.   **PARTIES ENTERING AFTER OCTOBER 1, 1997.** Any party which appears

25   after October 1, 1997, shall serve on all parties a "Notice of Appearance" identifying the party and

26   its counsel. Said party shall also comply with the terms herein within the time limits specified

27   in each appropriate paragraph or subparagraph, upon service of a conformed copy of this Order. Any

28   party who serves a summons and complaint or cross-complaint upon a new party after execution of

1   this Order shall serve a copy of plaintiffs' operative complaint, this Order, and all subsequent case

2   management orders upon the new party along with the summons.

3   3.   **CROSS-COMPLAINTS.** All parties who wish to pursue equitable indemnity/

4   contribution/declaratory relief causes of action against any party(ies) shall, within sixty (60) days

5   of appearing, file and serve a formal cross-complaint. Sixty (60) days after a party's first appearance,

6   leave of Court must be obtained. A cross-complaint seeking equitable indemnity/contribution only

7   shall not exceed three (3) pages. Counsel are encouraged, but not required, to use Judicial Council

8   Form 928.1(14). Cross-complaints adding any other theory of liability, alleging any new cause of

9   action, or adding a new party, shall be an exception to this Order and shall be filed, served and

10   responded to. Any complaint or cross-complaint served on a new party shall be accompanied with

11   all prior orders of the court, a copy of the most current service list, and the most current list of

12   construction defects. A formal notice of appearance shall be filed. The answer, with all applicable

13   affirmative defenses, shall be deemed filed.

14   Pursuant to *Code of Civil Procedure* section 411.35, complaints or cross-complaints

15   against professionals for whom a certificate of merit is required pursuant to said section shall be

16   accompanied by a certificate of merit by each party asserting any such complaint or cross-complaint.

17   All complaints or cross-complaints filed and served against any and all professionals for whom a

18   certificate of merit is required under *Code of Civil Procedure* section 411.35 must be answered

19   separately by said professional and any and all affirmative defenses must be raised in said pleadings

20   in the normal manner.

21   4.   **DESIGNATION AND FEES FOR MEDIATOR.** The parties hereto agree upon the

22   appointment of the Honorable _____(Ret.), ____address_____, per *Code of Civil Procedure*

23   sections 638 and 639(c) to act as a mediator for settlement conferences in this case and to hear

24   discovery disputes and make recommendations to court thereon. The special master shall preside

25   over mediation conferences, including the making of orders governing the attendance of parties,

26   counsel and insurance representatives to attend same. Except as to matters involving only a limited

27   number of parties (i.e., mediation or discovery disputes between two parties), the fees of the special

28   master conducting the mediation and all matters in accordance with this Case Management Order

-2-

1   shall be apportioned as follows: one-third (1/3) to plaintiffs; one-third (1/3) to direct defendants

2   apportioned   fifty percent (50%) to____general contractor____ and fifty percent (50%) to

3   ___subcontractor____; and one-third (1/3) to all remaining parties.

4   5.   **DISCOVERY.** Formal discovery between the parties is herein stayed other than as

5   set forth in this order until this order is hereafter modified.

6   6.   **NON-PARTY DISCOVERY.** All parties are allowed to conduct non-party document

7   discovery upon proper notice to all parties, and are required to deposit any such document discovery

8   in the depository, as set forth in the procedures below. Any party objecting to the non-party

9   discovery must seek a protective order and shall have the burden of establishing good cause for

10   prohibiting the service of such discovery.

11   7.   **INTERROGATORIES.** Defendants/cross-defendants shall serve verified responses

12   to the special interrogatories attached hereto as Exhibit A by October 1, 1997, or within thirty (30)

13   days after they appear, which ever is later, on all parties to this action.

14   8.   **STATEMENT OF WORK.** Defendants/cross-defendants are required to serve on

15   all parties a Statement of Work and/or involvement with the subject project, detailing the work done

16   by said party, by completing the Statement of Work attached hereto as Exhibit B by October 1, 1997,

17   or within thirty (30) days of the appearance of any party subsequent to the date of this order,

18   whichever is later.

19   9.   **DOCUMENT DEPOSITORIES.**

20   (a)   Location. All parties shall deposit all relevant, non-privileged documents as

21   described in Exhibits C and D attached hereto by October 1, 1997, or within thirty (30) days of the

22   appearance of any party subsequent to the date of this order, whichever is later. The depository shall

23   be located at _____. Any party wishing to inspect any documents therein shall contact the

24   depository coordinator to arrange for an appointment to do so. Parties shall be allowed access to the

25   depository within two business days of the request or as otherwise arranged between the party and

26   depository coordinator.

27   (b)   Index and Compliance. The deposit shall be accompanied by (1) an "Index of

28   Documents" deposited and (2) a "Notice of Compliance," which will also be served on all parties,

1   stating a clear and meaningful description of the document(s) produced, their bates-stamp numbers,

2   and the dates of production. Each document shall be consecutively bates-stamped with a number

3   and code designation as assigned in Exhibit E, and shall be bound in the following manner: double

4   hole-punched at the top of each document and accu-fastened with a stiff backing. Any party not

5   listed on Exhibit E should contact the depository coordinator for a bates-stamp assignment.

6   (c)   Privilege Log/Non-Produced Documents. Any party not depositing all

7   documents in its possession, custody or control shall, in the Notice of Compliance, (1) prepare a

8   serve a privilege log and identify the document(s) withheld with sufficient particularity for a motion

9   to compel, and (2) state the basis for refusing to produce each document including that particular

10   privilege(s) or doctrine(s) upon which protection against disclosure is based.

11   (d)   Oversized and Color Documents. Parties are required to deposit full-size

12   copies of oversized documents and color reprints of color photographs and other color documents

13   in their possession in the depository. Only the final version of any plans or drawings shall be

14   deposited. All versions other than the final version of the plans or drawings shall be included on the

15   document index and shall be made available for inspection and copying by any party upon ten (10)

16   days' written notice.

17   (e)   Requirement to Comply. All parties are under a continuing obligation to

18   deposit all non-privileged documents discovered after the initial production.

19   (f)   Expert Documents. All parties shall also deposit in the depository their

20   experts' files, including any reports generated thereby, using the designated prefix as set forth in

21   Exhibit E and following the same procedures for depositing the party's documents. Excluding

22   plaintiffs' and defendant ____'s repair recommendations, all expert files must be deposited in the

23   depository five (5) business days prior to the commencement of said expert's deposition.

24   Photographs need not be deposited, but must be made available at the deposition and for review and

25   copying five (5) business days prior to the deposition.

26   (g)   Maintenance of Originals. All parties agree to maintain the original documents

27   in their possession and deposit only copies in the depository. However, if a party wishes to see any

28   original documents, it shall be allowed to do so upon reasonable notice to counsel.

-4-

1   (h)   <u>Plans</u>. Any plans deposited pursuant to Exhibits C or D should include any

2   amendments thereto. Each party will deposit the original, a blueprint, or a vellum of every different

3   set of plans in its possession into the depository. Any party desiring to copy plans shall make its

4   own arrangements to copy documents requested from the depository, with the requesting party

5   responsible for arranging and paying for said service.

6   (i)   <u>Cost of Compliance</u>. The expense of complying with this section shall be

7   borne by the party depositing documents. The fees for maintaining and utilizing the depository shall

8   be as set by the custodian of the depository. A fee schedule shall be prepared in writing by the

9   custodian of the depository and provided to all parties.

10   Parties who have deposited documents which do not conform to this Order will be

11   given forty-eight (48) hours' notice to rectify the non-conforming deposit. Failure to comply and/or

12   rectify the non-conforming deposit will serve as a basis for a motion to compel, and a party in

13   violation may be subject to an order to show cause and/or sanctions.

14   10.   **PLAINTIFFS' LIST OF DEFECTS.**

15   (a)   <u>Preliminary List</u>. On or before October 22, 1997, plaintiffs shall deposit and

16   serve on all parties a preliminary list and matrix of all claimed defects/damages. Plaintiffs' list shall

17   be descriptive in nature, provide the location of each alleged defect. Any such list and damages

18   matrix is without prejudice for subsequent amendment, shall have no evidentiary impact, and is

19   merely for informational purposes.

20   (b)   <u>Final List</u>. A final summary of defects shall be served by plaintiffs on all

21   parties and deposited into the depository no later than February 1, 1998, and shall include the

22   following:

23   (1)   a description of the nature of each alleged defect;

24   (2)   the location of each alleged design or workmanship defect;

25   (3)   plaintiffs' contentions as to the cause of each alleged defect;

26   (4)   a description of each proposed repair of the alleged defect;

27   (5)   A remedial cost estimate depicting hard and soft costs as to each

28   alleged defect;

-5-

1   (6)   A description of and cost of any and all repairs undertaken to date as

2 to each alleged defect or, in the alternative, production into the depository of all repair invoices,

3 repair estimates, and checks related to said repairs;

4   (7)   A description of destructive testing done to date or, in the alternative,

5 production into the depository of all field notes, pictures, drawings, audio and visual recordings, and

6 reports relating to the testing.

7   The attachment of any consultant reports shall not constitute a waiver of any rights or

8 privileges afforded pursuant to Code of Civil Procedure §§ 2018 and 2034, *et seq.* Any defects/

9 damages discovered after this date shall only be allowed by leave of court.

10   If a new defect is discovered pursuant to the preparation of the final defect list,

11 plaintiffs shall afford all defendants the opportunity to inspect and/or test said subsequently

12 discovered defect. Any defect subsequently discovered may be added to the final defect list and final

13 repair estimate by stipulation or upon an ex parte application to the court and for good cause shown,

14 such good cause to be determined by the court.

15   Plaintiffs' report shall be protected as a document prepared for mediation in

16 accordance with §§ 1152 and 1152.5 of the Evidence Code.

17   (c)   <u>Augmentation of Final Defect List</u>. If, after February 1, 1998, plaintiffs desire

18 to augment the final defect list, a noticed motion must be brought to amend same. The motion will

19 address why the defect could not have been made part of the final defect list or discovered through

20 reasonable investigation on a previous date, and shall address the following for a showing of good

21 cause;

22   (1)   Proximity of the noticed motion to the trial date;

23   (2)   Whether allowing plaintiffs to augment will result in the inclusion of

24 new parties;

25   (3)   Whether augmentation list will necessitate additional destructive testing

26 or other investigation;

27   (4)   The likelihood of the augmentation resulting in a continuance of the

28 trial date;

-6-

1   (5) The burden on the defendants in having to conduct additional discovery

2   or investigation as a result of augmentation;

3   (6) Whether augmentation will require designation of additional experts.

4   If augmentation is allowed, all parties shall be afforded the opportunity to

5   timely investigate and inspect any such claimed defect/damage.

6   11. **PLAINTIFFS' COST TO REPAIR.**

7   (a) Preliminary list. On or before November 15, 1997, plaintiffs shall deposit and

8   serve on all parties a preliminary cost of repair estimate. Plaintiffs' estimate shall be descriptive in

9   nature, and describe the proposed repair for each of the defects outlined in plaintiffs' list of defects.

10   Any such cost of repair is without prejudice for a subsequent amendment, shall have no evidentiary

11   impact, and is merely for information purposes;

12   (b) Final cost of repair. A final cost of repair of defects shall be served by

13   plaintiffs on all parties and deposited into the depository no later than February 1, 1998.

14   Plaintiffs' cost of repair shall be protected as the document prepared for

15   mediation in accordance with Section 1152 and 1152.5 of the Evidence Code.

16   12. **NON-INTRUSIVE SITE INSPECTIONS.** Defendants and cross-defendants shall

17   serve on plaintiffs no later than September 19, 1998, a request to perform non-intrusive site

18   inspections, generally identifying the area(s) of said inspections, as depicted on Exhibit G. Plaintiffs

19   will serve all parties with a schedule of said non-intrusive inspections by September 26, 1997. Non-

20   intrusive site inspections will take place on October 7, 1997 through October 8, 1997.

21   13. **PRESENTATION OF DEFECTS/PERIPHERAL PARTY MEDIATION.** A

22   defect presentation will not be protected by *Evidence Code* section 1152 and will be held on

23   December 7, 1997, at which time plaintiffs will provide a presentation of all areas of defects and

24   deficiencies by their consultants/experts. The presentation will commence at 10:00 a.m. at the Law

25   Offices of _____.

26   14. **REPAIRS.** No significant repairs shall be performed without forty-eight (48) hours'

27   notice to Defendants. Emergency repairs may be performed with reasonable notice.

28   15. **DESTRUCTIVE TESTING.**

-7-

1   (a)   <u>By Plaintiffs</u>. Plaintiffs will give 5 days notice of any destructive testing they

2 conduct. All parties are entitled to observe plaintiffs' destructive testing. Any party, however, who

3 participates by sampling or causing the removal of any construction material during plaintiffs'

4 testing shall bear a pro rata share of the costs of conducting such testing. Any disputes over charges

5 relating to testing will be resolved at mediation. All testing conducted by plaintiffs must be

6 completed by December 1, 1997.

7   (b)   <u>By Defendants and Cross-defendants</u>.

8   (1)   All defendants and cross-defendants, and their consultants, will have

9 reasonable access to the property for the purpose of testing to analyze and evaluate plaintiffs' claims.

10 Such testing shall occur Monday-Friday between 9:00 a.m. and 4:30 p.m. Plaintiffs' counsel shall

11 use their best efforts to obtain all necessary cooperation regarding testing of the property. No

12 resident interviews will be permitted. Plaintiffs have the right to observe all testing conducted by

13 defendants/ cross-defendants. To the greatest extent possible, the defense shall coordinate

14 inspections to avoid redundancy and inconvenience to the residents.

15   (2)   Defendants/cross-defendants shall not have the right to retest locations

16 previously tested by another party if such defendant/cross-defendant was provided with seven (7)

17 days prior notice of the previous testing, except upon agreement by plaintiffs. In no event shall

18 destructive testing occur without forty-eight (48) hours prior notice to all parties.

19   (3)   Each party desiring to test shall submit a request setting forth the scope

20 and nature of its proposed testing to plaintiffs and defendants no later than November 15, 1997, on

21 the form attached as Exhibit F hereto. Plaintiffs and Defendants shall then meet and confer and

22 establish dates for the testing.

23   (4)   All repairs for destructive testing shall be done in a workmanlike

24 manner, in accordance with applicable codes, regulations and industry standards and practices. The

25 testing party shall also protect the property from the effects of inclement weather during destructive

26 testing and until it is restored to its previous condition.

27   All parties are entitled to observe all destructive testing. However, any

28 party who participates by sampling or causing the removal of any construction material shall bear

-8-

1   a pro rata share of the costs of conducting such destructive testing. Any disputes over charges
2   relating to destructive testing will be resolved at mediation.

3   (5)   Destructive testing by defendants/cross-defendants shall take place
4   December 1-7, 1997.

5   16.   **MEDIATION.** An initial mediation between Plaintiffs and Defendants will take place
6   on January 15-18, 1998, before The Honorable _____, at the offices of _____. Mediations will be
7   scheduled for the week of March 1, 1998, for all parties. Additional mediation may be scheduled
8   as necessary. Carriers or principals with full settlement authority are required to attend mediation,
9   unless notified otherwise.

10   Payment for mediation is to be one-third by plaintiffs, one-third by Defendants
11   allocated 50-50 between defendants _____, and one-third by cross-defendants when all parties are
12   involved, on a pro-rata basis, and one-half by plaintiffs and one-half by defendants, when only those
13   two parties are involved.

14   17.   **PLAINTIFFS' SETTLEMENT DEMAND.** Plaintiffs shall serve a settlement
15   demand on defendants, broken down into categories as to the nature of the alleged defects on or
16   before February 10, 1998.

17   18.   **DEFENDANTS' SETTLEMENT DEMAND.** Defendants shall serve a settlement
18   demand, broken down into categories by the nature of the alleged defect on cross-defendants on or
19   before February 25, 1998.

20   19.   **SETTLEMENT OFFERS.** Plaintiffs shall present any settlement offers made in
21   writing by any party in the action to defendants within five (5) days of receipt of same. Plaintiffs
22   shall transmit in writing a response to any such settlement offer within ten (10) days of receipt.

23   20.   **EXCHANGE OF EXPERT DESIGNATIONS.** The first expert exchange shall
24   occur on or about November 15, 1997, or within thirty (30) days after first appearing in this action,
25   whichever is later. A supplemental expert exchange shall may be made December 3, 1997.

26   21.   **DEPOSITION MEET AND CONFER.** Plaintiffs and Defendants will schedule a
27   meeting to establish protocol and a schedule for depositions of all experts, as appropriate. Said
28   meeting will take place prior to February 3, 1998. The depositions will take place between March

1 5, 1998 through April 30, 1998. Depositions of plaintiffs and party PMKs may be taken at anytime

2 pursuant to agreement of the parties as to time and place.

3   22.   **MOTION/DISCOVERY CUT-OFF DATES.** The last day for motions to be heard

4 will be April 17, 1998. All discovery shall be completed by April 17, 1998.

5   23.   **DISPOSITION CONFERENCE.** The disposition conference shall be set for April

6 23, 1998.

7   24.   **TRIAL.** Trial shall be set for May 8, 1998.

8   25.   **SANCTIONS.** Sanctions may be sought, subject to proper notice, upon failure to

9 comply with any of the provisions contained herein.

10   IT IS SO ORDERED.

11

12 DATE: _____   _____
                                            JUDGE OF THE SUPERIOR COURT

13

14

15

16

17

18

19

20

21

22

23

24

25

26

27

28

EXHIBIT A
SPECIAL INTERROGATORIES

<u>DEFINITIONS</u>

The term "policy of insurance" refers to any agreement under which any insurance carrier may be liable to satisfy, in whole or in part, a judgment that may be entered in this action, or to indemnify or reimburse for payments made to satisfy the judgment.

The term "damages" shall mean any actual or alleged weakness, fault, flaw, blemish, incomplete work, leak or condition causing any form of water infiltration or any construction condition indicating a failure to comply with the applicable plans or specifications, or a failure to comply with any applicable standard in the construction industry.

The term the "residence" means the single-family home which is the subject of this litigation.

<u>INTERROGATORIES</u>

1.   Are you a corporation? If so, state:

    a.   The names stated in the current articles of incorporation;

    b.   All other names used by the corporation during the past ten years and the dates each was used;

    c.   The date and place of incorporation;

    d.   The address of the principal place of business;

    e.   Whether you are qualified to do business in California.

2.   Are you a partnership? If so, state:

    a.   The current partnership name;

    b.   All other names used by the partnership during the past ten years and the dates each was used;

    c.   Whether you are a limited partnership and, if so, under the laws of what jurisdiction;

    d.   The name and address of each general partner;

    e.   The address of the principal place of business.

///

-11-

3. Are you a joint venture? If so, state:

    a. The current joint venture name;

    b. All other names used by the joint venture during the past ten years and the dates each was used;

    c. The name and address of each joint venturer;

    d. The address of the principal place of business.

4. Are you an unincorporated association? If so, state:

    a. The current unincorporated association name;

    b. All other names used by the unincorporated association during the past ten years and the dates each was used;

    c. The address of the principal place of business.

5. Have you done business under a fictitious name during the past ten years? If so, for each fictitious name, state:

    a. The name;

    b. The dates each was used;

    c. The state and county of each fictitious name filing;

    d. The address of the principal place of business.

6. Within the past five years, has any public entity registered or licensed your businesses? If so, for each license or registration:

    a. Identify the license or registration;

    b. State the name of the public entity;

    c. State the dates of issuance and expiration.

7. At the time of the acts alleged in the complaint, was there in effect any policy of insurance through which you were or might be insured in any manner (for example, primary, pro rata, or excess liability coverage) for the damages, claims or actions that have arisen out of the damages at the residence? If so, for each policy, state the kind of coverage. Please attach a copy of the policy or policies.

///

-12-

1   8.   For each policy listed in the responses to these interrogatories, please state the name
2   and address of the insurance company.
3   9.   For each policy listed in the responses to these interrogatories, please state the name
4   and address and telephone number of each named insured.
5   10.   For each policy listed in the responses to these interrogatories, please state the policy
6   number.
7   11.   For each policy listed in the responses to these interrogatories, please state the nature
8   of the limits of coverage for each type of coverage contained in the policy.
9   12.   For each policy listed in the responses to these interrogatories, please state whether
10   any reservation of rights or controversies or coverage disputes exist between you and the insurance
11   company. Please attach a copy of any reservation of rights letters.
12   13.   For each policy listed in the responses to these interrogatories, please state whether
13   any payments have been made under the policies. If so, state the policy and the amount of each
14   payment.
15   14.   For each policy listed in the responses to these interrogatories, please state the amount
16   of remaining coverage.
17   15.   For each policy listed in the responses to these interrogatories, please state the name,
18   address and telephone number of the custodian of the policy.
19   16.   Are you self-insured under any statute for the damages, claims or actions that have
20   arisen out of the damages at the residence? If so, specify the statute.
21   17.   For each policy listed in the responses to these interrogatories, please state whether
22   coverage is broad form coverage.
23   18.   Please provide information regarding the scope of the work you performed or services
24   you performed at the residence, including the phase number(s) and addresses.
25   19.   Please describe with specificity the work and/or services you performed at the
26   residence.
27   20.   Please provide the name, address and telephone number of the person most
28   knowledgeable regarding the work and/or services you performed at the residence.

-13-

1

2   **EXHIBIT B**

3   **STATEMENT OF WORK**

3   1.   NAME OF PARTY: _____

4   2.   NAME OF TRIAL ATTORNEY: _____

5   3.   DESCRIPTION OF WORK PERFORMED: _____

6   _____

7   4.   LOCATION OF WORK PERFORMED: _____

8   5.   INCLUSIVE DATES BETWEEN WHICH WORK WAS PERFORMED:

9   _____

10   6.   IDENTITY OF PERSON OR ENTITY WITH WHOM YOU CONTRACTED TO

11   PERFORM THE ABOVE-DESCRIBED WORK: _____

12   _____

13   7.   DID YOU SUPPLY MATERIALS? Yes _____ No _____

14   8.   IF YOU SUPPLIED MATERIALS, DESCRIBE THE MATERIALS YOU

15   PROVIDED: _____

16   9.   IF YOU SUPPLIED MATERIAL, IDENTIFY THE PERSON OR ENTITY FROM

17   WHOM YOU PURCHASED THE MATERIALS:

18   Name: _____

19   Address: _____

20   Telephone Number: _____

21   10.   DID YOU SUBCONTRACT ANY OF THE WORK THAT WAS TO BE

22   PERFORMED BY YOU TO ANOTHER PERSON OR ENTITY: Yes __ No __

23   11.   IF YOU DID SUBCONTRACT ANY OF YOUR WORK TO ANOTHER,

24   IDENTIFY THE PERSON OR ENTITY TO WHOM YOU SUBCONTRACTED:

25   _____

26   12.   IF YOU SUBCONTRACTED ANY OF YOUR WORK TO ANOTHER, WAS

27   THAT SUBCONTRACT IN WRITING? _____

28

-14-

1

2

3

EXHIBIT C

DESCRIPTION OF DOCUMENTS TO BE DEPOSITED
APPLICABLE TO PLAINTIFFS

4   1.   Any and all documents as defined in Evidence Code § 250, including but not limited

5   to, homeowner complaints, purchase and sale documents, memoranda, correspondence from or to

6   plaintiffs by anyone relating to any claimed defect/damage, offers to sell, listing agreements, rental

7   agreements, appraisals, refinancing documents and applications pertaining to same, photographs,

8   videos, documentation pertaining to any improvements to the property, employment contracts,

9   claims made to any insurance company, disclosure made and/or received with respect to purchase

10   and/or sale of the property, consultant reports, or any other documentation that would reflect or

11   confirm dates of manifestation or observation of the claimed defects/damages.

12   2.   Final defects listed as required by this order.

13   3.   All escrow documents, homeowner disclosure forms, real estate agent/broker

14   disclosure forms (seller and buyer), all modifications and/or addenda to all escrow and disclosure

15   forms, real estate purchase contracts and addenda, all offers and acceptances, escrow instructions,

16   inspection reports, multiple listing forms, and all advertising.

17   4.   Any and all expert documents as required by this order.

18   5.   Any and all documents subpoenaed from third parties as required by this order.

19

20

21

22

23

24

25

26

27

28

1
2
3
4
5
6
7
8
9
10
11
12
13
14
15
16
17
18
19
20
21
22
23
24
25
26
27
28

EXHIBIT D

DESCRIPTION OF DOCUMENTS TO BE DEPOSITED
APPLICABLE TO ALL PARTIES EXCEPT PLAINTIFFS

1.   Any and all documents as defined in Evidence Code §250, including but not limited to any and all contracts, subcontracts, sub-subcontracts, agreements, job files, blueprints, plans, specifications, notes, memoranda, advertisement, correspondence, photographs, diagrams, calculations, invoices, purchase orders, change orders, addenda, reports, journals, marketing documents, job diaries, receipts, accounting records, writings, amendments to all plans, governmental inspector punch lists and sign-off sheets and/or other documents including discovery from previous litigation, referring to and/or concerning the development, construction and/or repair of plaintiffs' property.

2.   Any and all insurance policies which may potentially provide insurance coverage for any claim asserted against any party in this lawsuit regardless of whether coverage has been asserted to be inapplicable or denied by any insurance company.

3.   Any and all reservation-of-rights letters received from insurance carriers pertaining to this matter.

4.   All escrow documents, homeowner disclosure forms, real estate agent/broker disclosure forms (seller and buyer), all modifications and/or addenda to all escrow and disclosure forms, real estate purchase contracts and addenda, all offers and acceptances, escrow instructions, inspection reports, multiple listing forms, and all advertising.

5.   Any and all expert documents as required by this order.

6.   Any and all documents subpoenaed from third parties as required by this order.

1

2

3

4

5

6

7

8

9

10

11

12

13

14

15

16

17

18

19

20

21

22

23

24

25

26

27

28

EXHIBIT E

BATES STAMP PREFIX CODES

PARTY	CODE
Plaintiffs	PL-
Developer	DV-

1

2

3

4

5

6

7

8

9

10

11

12

13

14

15

16

17

18

19

20

21

22

23

24

25

26

27

28

EXHIBIT F

DESTRUCTIVE TESTING REQUEST

NAME OF ATTORNEY: _____

NAME OF PARTY: _____

AREAS OF WORK PERFORMED BY YOUR CLIENT: _____

_____

LOCATIONS YOU WOULD LIKE TO HAVE DESTRUCTIVELY TESTED: _____

_____

_____

_____

_____

TYPE OF DESTRUCTIVE TESTING CONTEMPLATED: _____

_____

_____

_____

_____

IS SPECIAL EQUIPMENT REQUIRED FOR TESTING: YES ☐   NO ☐

IF YES, PLEASE LIST TYPE OF EQUIPMENT NECESSARY: _____

_____

_____

ESTIMATED TIME NEEDED PER LOCATION: _____

ARE YOU WILLING TO SHARE IN THE COSTS OF DESTRUCTIVE TESTING?

        YES ☐   NO ☐

DATED: _____   BY:_____

1	**EXHIBIT G**
2	**NON-INTRUSIVE INSPECTION REQUEST**
3	
4	NAME OF ATTORNEY: _____
5	
6	NAME OF PARTY: _____
7	
8	AREAS OF WORK PERFORMED BY YOUR CLIENT: _____
9	_____
10	
11	LOCATIONS YOU WOULD LIKE TO INSPECT: _____
12	_____
13	_____
14	_____
15	_____
16	
17	ESTIMATED TIME NEEDED PER LOCATION: _____
18	
19	DATED: _____ BY:_____
20	
21	
22	
23	
24	
25	
26	
27	
28	

EXHIBIT H

SUMMARY OF SIGNIFICANT DEADLINES

Comply with Case Management Order	October 1, 1997, or 30 days after appearance, whichever is later
File Equitable Cross-complaint	October 1, 1997, or within 60 days of filing answer/notice of appearance
Plaintiffs' List of Defects - Preliminary	October 22, 1997
Plaintiffs' Cost of Repair - Preliminary	November 15, 1997
Non-Intrusive Site Inspection Requests	September 19, 1997
Add New Parties	October 1, 1997
Non-Intrusive Site Inspection Schedule	September 26, 1997
Non-Intrusive Site Inspections	October 7-8, 1997
Completion of Plaintiffs' Destructive Testing	October 31, 1997
Plaintiffs' List of Defects and Cost to Repair -- Final	February 1, 1998
Plaintiffs' Presentation of Defects and Peripheral Party Mediation	December 7, 1997
Defense Destructive Testing Requests	November 15, 1997
Defense Destructive Testing	December 1-5, 1997
First Expert Designation	November 15, 1997, or within 30 days of appearing in action, whichever is later
Supplemental Expert Designation	December 3, 1997
Experts Meetings	To Be Determined
Mediations (Plaintiffs and Developer)	January 15-18, 1998
Plaintiffs' Settlement Demand	February 10, 1998
Defendants' Settlement Demand	February 25, 1998
Mediation (all parties)	March 1, 1998
Depositions (Experts)	March 5-April 30, 1998
Motion/Discovery Cut-Off	April 17, 1998
Disposition/Pre-Trial Conference	April 23, 1998
Trial	May 8, 1998

ACI (1982). *Guide to Durable Concrete.* ACI Committee 201, American Concrete Institute, Detroit.

ACI (1990). *ACI Manual of Concrete Practice, Part 1, Materials and General Properties of Concrete.* American Concrete Institute, Detroit.

ASCE (1964). *Design of Foundations for Control of Settlement.* American Society of Civil Engineers, New York, 592 pp.

ASCE (1972). "Subsurface Investigation for Design and Construction of Foundations of Buildings." Task Committee for Foundation Design Manual. Part I, *Journal of Soil Mechanics,* ASCE, vol. 98, no. SM5, pp. 481–490; Part II, no. SM6, pp. 557–578; Parts III and IV, no. SM7, pp. 749–764.

ASCE (1976). *Subsurface Investigation for Design and Construction of Foundations of Buildings.* Manual No. 56. American Society of Civil Engineers, New York, 61 pp.

ASCE (1978). *Site Characterization and Exploration.* Proceedings of the Specialty Workshop at Northwestern University, C. H. Dowding, ed. New York, 395 pp.

ASCE (1987). *Civil Engineering Magazine,* American Society of Civil Engineers, New York, April.

ASFE (undated). *Expert: A Guide to Forensic Engineering and Service as an Expert Witness.* ASFE: The Association of Engineering Firms Practicing in the Geosciences, Silver Spring, Md., 52 pp.

ASFE (1993). *Recommended Practices for Design Professionals Engaged as Experts in the Resolution of Construction Industry Disputes.* ASFE: The Association of Engineering Firms Practicing in the Geosciences, Silver Spring, Md., 8 pp.

ASTM (1970). "Special Procedures for Testing Soil and Rock for Engineering Purposes." ASTM Special Technical Publication 479, Philadelphia, 630 pp.

ASTM (1971). "Sampling of Soil and Rock." ASTM Special Technical Publication 483, Philadelphia, 193 pp.

ASTM (1997a). *Annual Book of ASTM Standards: Concrete and Aggregates,* vol. 04.2. Standard No. C 881-90, "Standard Specification for Epoxy-Resin-Base Bonding Systems for Concrete," West Conshohocken, Pa., pp. 436–440.

ASTM (1997b). *Annual Book of ASTM Standards: Road and Paving Materials; Vehicle-Pavement Systems,* vol. 04.03. Standard No. D 5340-93, "Standard Test Method for Airport Pavement Condition Index Surveys," West Conshohocken, Pa., pp. 546–593.

ASTM (1997c). *Annual Book of ASTM Standards: Road and Paving Materials; Vehicle-Pavement Systems,* vol. 04.03. Standard No. E 1778-96a, "Standard Terminology Relating to Pavement Distress," West Conshohocken, Pa., pp. 843–849.

ASTM (1997d). *Annual Book of ASTM Standards,* vol. 04.08, *Soil and Rock (I).* Standard No. D 420-93, "Standard Guide to Site Characterization for Engineering, Design, and Construction Purposes," West Conshohocken, Pa., pp. 1–7.

ASTM (1997e). *Annual Book of ASTM Standards,* vol. 04.08, *Soil and Rock (I).* Standard No. D 422-90, "Standard Test Method for Particle-Size Analysis of Soils," West Conshohocken, Pa., pp. 10–20.

ASTM (1997f). *Annual Book of ASTM Standards,* vol. 04.08, *Soil and Rock (I).* Standard No. D 698-91, "Test Method for Laboratory Compaction Characteristics of Soil Using Standard Effort," West Conshohocken, Pa., pp. 77–87.

ASTM (1997g). *Annual Book of ASTM Standards,* vol. 04.08, *Soil and Rock (I).* Standard No. D 1557-91, "Test Method for Laboratory Compaction Characteristics of Soil Using Modified Effort," West Conshohocken, Pa., pp. 126–133.

ASTM (1997h). *Annual Book of ASTM Standards,* vol. 04.08, *Soil and Rock (I).* Standard No. D 2435-96, "Standard Test Method for One-Dimensional Consolidation Properties of Soils," West Conshohocken, Pa., pp. 207–216.

ASTM (1997i). *Annual Book of ASTM Standards,* vol. 04.08, *Soil and Rock (I).* Standard No. D 2844-94, "Standard Test Method for Resistance R-Value and Expansion Pressure of Compacted Soils," West Conshohocken, Pa., pp. 246–253.

ASTM (1997j). *Annual Book of ASTM Standards,* vol. 04.08, *Soil and Rock (I).* Standard No. D 2974-95, "Standard Test Methods for Moisture, Ash, and Organic Matter of Peat and Other Organic Soils," West Conshohocken, Pa., pp. 285–287.

ASTM (1997k). *Annual Book of ASTM Standards,* vol. 04.08, *Soil and Rock (I).* Standard No. D 4829-95, "Standard Test Method for Expansion Index of Soils," West Conshohocken, Pa., pp. 866–869.

ASTM (1997l). *Annual Book of ASTM Standards,* vol. 04.09, *Soil and Rock (II), Geosynthetics.* Standard No. D 5333-92, "Standard Test Method for Measurement of Collapse Potential of Soils," West Conshohocken, Pa., pp. 225–227.

Abramson, L. W., Lee, T. S., Sharma, S., and Boyce, G. M. (1996). *Slope Stability and Stabilization Methods.* John Wiley & Sons, New York, 629 pp.

"Accident Investigation Report" (no date). Cal/OSHA Report No. 018, California Occupational Safety and Health Administration, San Francisco, 1 p.

"Aggressive Chemical Exposure" (1990). *ACI Manual of Concrete Practice, Part 1, Materials and General Properties of Concrete,* American Concrete Institute, Detroit, pp. 201.2R-10 to 201.2R-13.

Al-Homoud, A. S., Basma, A. A., Husein Malkawi, A. I., and Al Bashabsheh, M. A. (1995). "Cyclic Swelling Behavior of Clays." *Journal of Geotechnical Engineering,* ASCE, vol. 121, no. 7, pp. 562–565.

Al-Homoud, A. S., Basma, A. A., Husein Malkawi, A. I., and Al Bashabsheh, M. A. (1997). Closure of "Cyclic Swelling Behavior of Clays." *Journal of Geotechnical and Geoenvironmental Engineering,* ASCE, vol. 123, no. 8, pp. 786–788.

Al-Khafaji, A. W. N., and Andersland, O. B. (1981). "Ignition Test for Soil Organic Content Measurement." *Journal of Geotechnical Engineering Division,* ASCE, vol. 107, no. 4, pp. 465–479.

Alvarado Soils Engineering (1977). "Preliminary Geologic and Soils Investigation, Proposed 4 Lot Residential Development, Southeast Side of Desert View Drive." Project No. 52C1E, San Diego, 26 pp.

Ambraseys, N. N. (1960). "On the Seismic Behavior of Earth Dams." *Proceedings of the Second World Conference on Earthquake Engineering,* vol. 1, Tokyo and Kyoto, pp. 331–358.

Anderson, S. A., and Sitar, N. (1995). "Analysis of Rainfall-Induced Debris Flow." *Journal of Geotechnical Engineering,* ASCE, vol. 121, no. 7, pp. 544–552.

Anderson, S. A., and Sitar, N. (1996). Closure of "Analysis of Rainfall-Induced Debris Flow." *Journal of Geotechnical Engineering,* ASCE, vol. 122, no. 12, pp. 1025–1027.

Association of Engineering Geologists (1978). *Failure of St. Francis Dam.* Southern California Section.

Athanasopoulos, G. A. (1995). Discussion of "1988 Armenia Earthquake II. Damage Statistics versus Geologic and Soil Profiles." *Journal of Geotechnical Engineering,* ASCE, vol. 121, no. 4, pp. 395–398.

Baldwin, J. E., Donley, H. F., and Howard, T. R. (1987). "On Debris Flow/Avalanche Mitigation and Control, San Francisco Bay Area, California." *Debris Flow/Avalanches: Process, Recognition, and Mitigation,* The Geological Society of America, Boulder, Colo. pp. 223–226.

Bates, R. L., and Jackson, J. A. (1980). *Glossary of Geology.* American Geological Institute, Falls Church, Va., 751 pp.

Bellport, B. P. (1968). "Combating Sulphate Attack on Concrete on Bureau of Reclamation Projects." *Performance of Concrete, Resistance of Concrete to Sulphate and Other Environmental Conditions,* University of Toronto Press, Toronto, pp. 77–92.

Benton Engineering Inc. (1962). "Landslide Investigations. Soledad Mountain, City of San Diego, California." Project No. 61-12-14F, prepared for City of San Diego, 25 pp.

Best, M. G. (1982). *Igneous and Metamorphic Petrology.* W. H. Freeman, San Francisco.

Biddle, P. G. (1979). "Tree Root Damage to Buildings—An Arboriculturist's Experience." *Arboricultural Journal,* vol. 3, no. 6, pp. 397–412.

Biddle, P. G. (1983). "Patterns of Soil Drying and Moisture Deficit in the Vicinity of Trees on Clay Soils." *Geotechnique,* London, vol. 33, no. 2, pp. 107–126.

Bishop, A. W. (1955). "The Use of the Slip Circle in the Stability Analysis of Slopes." *Geotechnique,* London, vol. 5, no. 1, pp. 7–17.

Bishop, A. W., and Henkel, D. J. (1962). *The Measurement of Soil Properties in the Triaxial Test.* 2d ed., Edward Arnold, London, 228 pp.

Bjerrum, L. (1963). "Allowable Settlements of Structures." *Proceedings of European Conference on Soil Mechanics and Foundation Engineering,* vol. 2, Wiesbaden, Germany, pp. 135–137.

Boardman, B. T., and Daniel, D. E. (1996). "Hydraulic Conductivity of Desiccated Geosynethic Clay Liners." *Journal of Geotechnical Engineering,* ASCE, vol. 122, no. 3, pp. 204–215.

Bonilla, M. G. (1970). "Surface Faulting and Related Effects." Chapter 3 of *Earthquake Engineering,* Robert L. Wiegel, coordinating editor. Prentice-Hall, Englewood Cliffs, N.J., pp. 47–74.

Boone, S. T. (1996). "Ground-Movement-Related Building Damage." *Journal of Geotechnical Engineering,* ASCE, vol. 122, no. 11, pp. 886–896.

Boscardin, M. D., and Cording, E. J. (1989). "Building Response to Excavation-Induced Settlement." *Journal of Geotechnical Engineering,* ASCE, vol. 115, no. 1, pp. 1–21.

Bourdeaux, G., and Imaizumi, H. (1977). "Dispersive Clay at Sabradinho Dam." *Dispersive Clays, Related Piping, and Erosion in Geotechnical Projects,* STP 625, American Society for Testing and Materials, Philadelphia, pp. 12–24.

Bowles, J. E. (1982). *Foundation Analysis and Design,* 3d ed., McGraw-Hill, New York, 816 pp.

Brewer, H. W. (1965). "Moisture Migration—Concrete Slab-on-Ground Construction." *Journal of the PCA Research and Development Laboratories,* May, pp. 2–17.

Bromhead, E. N. (1984). *Ground Movements and their Effects on Structures,* Chap, 3, "Slopes and Embankments." P. B. Attewell and R. K. Taylor, eds., Surrey University Press, London, p. 63.

Brown, D. R., and Warner, J. (1973). "Compaction Grouting." *Journal of the Soil Mechanics and Foundations Division,* ASCE, vol. 99, no. SM8, pp. 589–601.

Brown, R. W. (1990). *Design and Repair of Residential and Light Commercial Foundations.* McGraw-Hill, New York, 241 pp.

Brown, R. W. (1992). *Foundation Behavior and Repair, Residential and Light Commercial.* McGraw-Hill, New York, 271 pp.

Bruce, D. A., and Jewell, R. A. (1987). "Soil Nailing: The Second Decade." *International Conference on Foundations and Tunnels,* London, England, pp. 68–83.

Burland, J. B., Broms, B. B., and DeMello, V. F. B. (1977). "Behavior of Foundations and Structures: State of the Art Report." *Proceedings of the 9th International Conference on Soil Mechanics and Foundation Engineering,* Japanese Geotechnical Society, Tokyo, pp. 495–546.

Butt, T. K. (1992). "Avoiding and Repairing Moisture Problems in Slabs on Grade." *The Construction Specifier,* December, pp. 17–27.

Byer, J. (1992). "Geocalamities—The Do's and Don'ts of Geologic Consulting in Southern California." *Engineering Geology Practice in Southern California.* B. W. Pipkin and R. J. Proctor, eds., Star Publishing, Association of Engineering Geologists, Southern California Section, Special Publication No. 4, pp. 327–337.

California Department of Water Resources (1967). "Earthquake Damage to Hydraulic Structures in California." California Department of Water Resources, Bulletin 116-3, Sacramento.

California Division of Highways (1973). *Flexible Pavement Structural Design Guide for California Cities and Counties,* Sacramento, 42 pp.

Carper, K. L. (1986). *Forensic Engineering: Learning from Failures.* ASCE, New York, 98 pp.

Carper, K. L. (1989). *Forensic Engineering.* Elsevier, New York, 361 pp.

Carpet and Rug Institute (1995). *Standard Industry Reference Guide for Installation of Residential Textile Floor Covering Materials,* 3d ed., Carpet and Rug Institute, Dalton, Ga., 44 pp.

Casagrande, A. (1932). Discussion of "A New Theory of Frost Heaving" by A. C. Benkelman and F. R. Ohlmstead, *Proceedings of the Highway Research Board,* vol. 11, pp. 168–172.

Casagrande, A. (1948). "Classification and Identification of Soils." *Transactions ASCE,* vol. 113, p. 901.

Cedergren, H. R. (1989). *Seepage, Drainage, and Flow Nets,* 3d ed. John Wiley & Sons, New York, 465 pp.

Cernica, J. N. (1995a). *Geotechnical Engineering: Soil Mechanics.* John Wiley & Sons, New York, 454 pp.

Cernica, J. N. (1995b). *Geotechnical Engineering: Foundation Design.* John Wiley & Sons, New York, 486 pp.

Cheeks, J. R. (1996). "Settlement of Shallow Foundations on Uncontrolled Mine Spoil Fill." *Journal of Performance of Constructed Facilities,* ASCE, vol. 10, no. 4, pp. 143–151.

Chen, F. H. (1988). *Foundations on Expansive Soil,* 2d ed., Elsevier, New York, 463 pp.

Cheney, J. E., and Burford, D. (1975). "Damaging Uplift to a Three-Story Office Block Constructed on a Clay Soil Following Removal of Trees." *Proceedings, Conference on Settlement of Structures, Cambridge,* Pentech Press, London, pp. 337–343.

"Citation" (1986). Identification No. W-5640, California Occupational Safety and Health Administration, San Francisco, September 4, 1 p.

Cleveland, G. B. (1960). "Geology of the Otay Clay Deposit, San Diego County, California." *California Division of Mines Special Report 64,* Sacramento, 16 pp.

Coduto, D. P. (1994). *Foundation Design, Principles and Practices.* Prentice Hall, Englewood Cliffs, N.J., 796 pp.

Collins, A. G., and Johnson, A. I. (1988). *Ground-Water Contamination, Field Methods,* symposium papers published by American Society for Testing and Materials, Philadelphia, 491 pp.

Committee Report for the State (1928). "Causes Leading to the Failure of the St. Francis Dam." California Printing Office.

Compton, R. R. (1962). *Manual of Field Geology.* John Wiley and Sons, New York, pp. 255–256.

Corns, C. F. (1974). "Inspection Guidelines—General Aspects." *Safety of Small Dams, Proceedings of the Engineering Foundation Conference,* Henniker, N.H., published by ASCE, New York, pp. 16–21.

Cutler, D. F., and Richardson, I. B. (1989). *Tree Roots and Buildings,* 2d ed. Longman, England, pp. 1–67.

David, D., and Komornik, A. (1980). "Stable Embedment Depth of Piles in Swelling Clays." *Fourth International Conference on Expansive Soils,* ASCE, vol. 2, Denver, pp. 798–814.

Day, R. W. (1989). "Relative Compaction of Fill Having Oversize Particles." *Journal of Geotechnical Engineering,* ASCE, vol. 115, no. 10, pp. 1487–1491.

Day, R. W. (1990a). "Differential Movement of Slab-on-Grade Structures." *Journal of Performance of Constructed Facilities,* ASCE, vol. 4, no. 4, pp. 236–241.

Day, R. W. (1990b). "Index Test for Erosion Potential." *Bulletin of the Association of Engineering Geologists,* vol. 27, no. 1, pp. 116–117.

Day, R. W. (1991a). Discussion of "Collapse of Compacted Clayey Sand." *Journal of Geotechnical Engineering,* ASCE, vol. 117, no. 11, pp. 1818–1821.

Day, R. W. (1991b). "Expansion of Compacted Gravelly Clay." *Journal of Geotechnical Engineering,* ASCE, vol. 117, no. 6, pp. 968–972.

Day, R. W. (1992a). "Effective Cohesion for Compacted Clay." *Journal of Geotechnical Engineering,* ASCE, vol. 118, no. 4, pp. 611–619.

Day, R. W. (1992b). "Swell Versus Saturation for Compacted Clay." *Journal of Geotechnical Engineering,* ASCE, vol. 118, no. 8, pp. 1272–1278.

Day, R. W. (1992c). "Walking of Flatwork on Expansive Soils." *Journal of Performance of Constructed Facilities,* ASCE, vol. 6, no. 1, pp. 52–57.

Day, R. W. (1992d). "Moisture Migration Through Concrete Floor Slabs." *Journal of Performance of Constructed Facilities,* ASCE, vol. 6, no. 1, pp. 46–51.

Day, R. W. (1992e). "Depositions and Trial Testimony, A Positive Experience?" *Journal of Professional Issues in Engineering Education and Practice,* ASCE, vol. 118, no. 2, pp. 129–131.

Day, R. W. (1993). "Surficial Slope Failure: A Case Study." *Journal of Performance of Constructed Facilities,* ASCE, vol. 7, no. 4, pp. 264–269.

Day, R. W. (1994a). Discussion of "Evaluation and Control of Collapsible Soil." *Journal of Geotechnical Engineering,* ASCE, vol. 120, no. 5, pp. 924–925.

Day, R. W. (1994b). "Performance of Slab-on-Grade Foundations on Expansive Soil." *Journal of Performance of Constructed Facilities,* ASCE, vol. 8, no. 2, pp. 129–138.

Day, R. W. (1994c). "Surficial Stability of Compacted Clay: Case Study." *Journal of Geotechnical Engineering,* ASCE, vol. 120, no. 11, pp. 1980–1990.

Day, R. W. (1994d). "Moisture Migration Through Basement Walls." *Journal of Performance of Constructed Facilities,* ASCE, vol. 8, no. 1, pp. 82–86.

Day, R. W. (1995a). "Pavement Deterioration: Case Study." *Journal of Performance of Constructed Facilities,* ASCE, vol. 9. no. 4, pp. 311–318.

Day, R. W. (1995b). "Reactivation of an Ancient Landslide." *Journal of Performance of Constructed Facilities,* ASCE, vol. 9, no. 1, pp. 49–56.

Day, R. W. (1995c). "Engineering Properties of Diatomaceous Fill." *Journal of Geotechnical Engineering,* ASCE, vol. 121, no. 12, pp. 908–910.

Day, R. W. (1996a). "Study of Capillary Rise and Thermal Osmosis." *Journal of Environmental and Engineering Geoscience,* joint publication, AEG and GSA, vol. 2, no. 2, pp. 249–254.

Day, R. W. (1996b). "Moisture Penetration of Concrete Floor Slabs, Basement Walls, and Flat Slab Ceilings." *Practice Periodical on Structural Design and Construction,* ASCE, vol. 1, no. 4, pp. 104–107.

Day, R. W. (1996c). "Repair of Damaged Slab-on-Grade Foundations." *Practice Periodical on Structural Design and Construction,* ASCE, vol. 1, no. 3, pp. 83–87.

Day, R. W. (1997a). "Soil Related Damage to Tilt-up Structures." *Practice Periodical on Structural Design and Construction,* ASCE, vol. 2, no. 2, pp. 55–60.

Day, R. W. (1997b). "Hydraulic Conductivity of a Desiccated Clay Upon Wetting." *Journal of Environmental and Engineering Geoscience,* Joint Publication, AEG and GSA, vol. 3, no. 2, pp. 308–311.

Day, R. W. (1997c). "Design and Construction of Cantilevered Retaining Walls." *Practice Periodical on Structural Design and Construction,* ASCE, vol. 2, no. 1, pp. 16–21.

Day, R. W. (1998a). Discussion of "Ground-Movement-Related Building Damage." *Journal of Geotechnical and Geoenvironmental Engineering,* ASCE, vol. 124, no. 5, pp. 462–465.

Day, R. W. (1998b). "Settlement Behavior of Post-Tensioned Slab-on-Grade." *Journal of Performance of Constructed Facilities,* ASCE, vol. 12, no. 2, pp. 56–61.

Day, R. W., and Axten, G. W. (1989). "Surficial Stability of Compacted Clay Slopes." *Journal of Geotechnical Engineering,* ASCE, vol. 115, no. 4, pp. 577–580.

Day, R. W., and Axten, G. W. (1990). "Softening of Fill Slopes Due to Moisture Infiltration." *Journal of Geotechnical Engineering,* ASCE, vol. 116, no. 9, pp. 1424–1427.

Day, R. W., and Poland, D. M. (1996). "Damage Due to Northridge Earthquake Induced Movement of Landslide Debris." *Journal of Performance of Constructed Facilities,* ASCE, vol. 10, no. 3, pp. 96–108.

Department of the Army (1970). *Engineering and Design, Laboratory Soils Testing,* Engineer Manual EM 1110-2-1906. Prepared at the U.S. Army Engineer Waterways

Experiment Station, published by the Department of the Army, Washington, D.C., 282 pp.

Design and Control of Concrete Mixtures (1988). 13th ed., Portland Cement Association, Stokie, Ill., 205 pp.

Diaz, C. F., Hadipriono, F. C., and Pasternack, S. (1994). "Failure of Residential Building Basements in Ohio." *Journal of Performance of Constructed Facilities,* ASCE, vol. 8, no. 1, pp. 65–80.

Driscol, R. (1983). "The Influence of Vegetation on the Swelling and Shrinkage Caused by Large Trees." *Geotechnique,* London, England, vol. 33, no. 2, pp. 1–67.

Dudley, J. H. (1970). "Review of Collapsing Soils," *Journal of Soil Mechanics and Foundation Engineering Division,* ASCE, vol. 96, no. SM3, pp. 925–947.

Duke, C. M. (1960). "Foundations and Earth Structures in Earthquakes." *Proceedings of the Second World Conference on Earthquake Engineering,* vol. 1, Tokyo and Kyoto, Japan, pp. 435–455.

Duncan, J. M. (1996). "State of the Art: Limit Equilibrium and Finite-Element Analysis of Slopes." *Journal of Geotechnical and Geoenvironmental Engineering,* ASCE, vol. 122, no. 7, pp. 577–596.

Duncan, J. M., Williams, G. W., Sehn, A. L., and Seed, R. B. (1991). "Estimation Earth Pressures due to Compaction." *Journal of Geotechnical Engineering,* ASCE, vol. 117, no. 12, 1833–1847.

Dyni, R. C., and Burnett, M. (1993). "Speedy Backfilling for Old Mines." *Civil Engineering Magazine,* ASCE, vol. 63, no. 9, pp. 56–58.

Earth Manual (1985). A Water Resources Technical Publication, 2d ed., U.S. Department of the Interior, Bureau of Reclamation, Denver, 810 pp.

Ehlig, P. L. (1986). "The Portuguese Bend Landslide: Its Mechanics and a Plan for its Stabilization." *Landslides and Landslide Mitigation in Southern California,* 82nd Annual Meeting of the Cordilleran Section of the Geological Society of America, Los Angeles, pp. 181–190.

Ehlig, P. L. (1992). "Evolution, Mechanics, and Migration of the Portuguese Bend Landslide, Palos Verdes Peninsula, California." *Engineering Geology Practice in Southern California,* B. W. Pipkin and R. J. Proctor, eds., Star Publishing, Association of Engineering Geologists, Southern California Section, special publication no. 4, pp. 531–553.

Ellen, S. D., and Fleming, R. W. (1987). "Mobilization of Debris Flows from Soil Slips, San Francisco Bay Region, California." *Debris Flows/Avalanches: Process, Recognition, and Mitigation,* The Geological Society of America, Boulder, Colo., pp. 31–40.

Engineering News-Record (1987). McGraw-Hill, New York, February 5.

Evans, D. A. (1972). *Slope Stability Report,* Slope Stability Committee, Department of Building and Safety, Los Angeles.

"Excavations, Final Rule" (1989). *29CFR Part 1926, Federal Register,* vol. 54, no. 209, pp. 45894–45991.

Fairweather, V. (1992). "L'Ambiance Plaza: What Have We Learned." *Civil Engineering Magazine,* ASCE, vol. 62, no. 2, pp. 38–41.

Feld, J. (1965). "Tolerance of Structures to Settlement." *Journal of Soil Mechanics,* ASCE, vol. 91, no. SM3, pp. 63–77.

Feld, J., and Carper, K. L. (1997). *Construction Failure.* 2d ed., John Wiley & Sons, New York, 512 pp.

FitzSimons, N. (1986). "An Historic Perspective of Failures of Civil Engineering Works." *Forensic Engineering, Learning from Failures.* ASCE, New York, pp. 38–45.

Florensov, N. A., and Solonenko, V. P., eds. (1963). "Gobi-Altayskoye Zemletryasenie." *Izvestiya Akademii Nauk SSSR.*; also 1965, *The Gobi-Altai Earthquake,* U.S. Department of Commerce (English translation), Washington, D.C.

Foshee, J., and Bixler, B. (1994). "Cover-Subsidence Sinkhole Evaluation of State Road 434, Longwood, Florida." *Journal of Geotechnical Engineering,* ASCE, vol. 120, no. 11, pp. 2026–2040.

Fourie, A. B. (1989). "Laboratory Evaluation of Lateral Swelling Pressures." *Journal of Geotechnical Engineering,* ASCE, vol. 115, no. 10, pp. 1481–1486.

Franklin, A. G., Orozco, L. F., and Semrau, R. (1973). "Compaction and Strength of Slightly Organic Soils." *Journal of Soil Mechanics and Foundations Division, ASCE,* vol. 99, no. 7, pp. 541–557.

Fredlund, D. G., and Rahardjo, H. (1993). *Soil Mechanics for Unsaturated Soil.* John Wiley & Sons, New York, 517 pp.

Geo-Slope (1991). *User's Guide, SLOPE/W for Slope Stability Analysis,* version 2, Geo-Slope International, Calgary, 444 pp.

Gill, L. D. (1967). "Landslides and Attendant Problems." *Mayor's Ad Hoc Landslide Committee Report,* Los Angeles.

Goh, A. T. C. (1993). "Behavior of Cantilever Retaining Walls." *Journal of Geotechnical Engineering, ASCE,* vol. 119, no. 11, 1751–1770.

Gould, J. P. (1995). "Geotechnology in Dispute Resolution," The Twenty-sixth Karl Terzaghi Lecture. *Journal of Geotechnical Engineering, ASCE,* vol. 121, no. 7, pp. 521–534.

Graf, E. D. (1969). "Compaction Grouting Techniques," *Journal of the Soil Mechanics and Foundations Division, ASCE,* vol. 95, no. SM5, pp. 1151–1158.

Grant, R., Christian, J. T., and Vanmarcke, E. H. (1974). "Differential Settlement of Buildings." *Journal of Geotechnical Engineering, ASCE,* vol. 100, no. 9, pp. 973–991.

Grantz, A., Plafker, G., and Kachadoorian, R. (1964). *Alaska's Good Friday Earthquake, March 27, 1994.* Department of the Interior, Geological Survey Circular 491, Washington, D.C.

Gray, R. E. (1988). "Coal Mine Subsidence and Structures." *Mine Induced Subsidence: Effects on Engineered Structures,* Geotechnical Special Publication 19, ASCE, New York, pp. 69–86.

Greenfield, S. J., and Shen, C. K. (1992). *Foundations in Problem Soils.* Prentice-Hall, Englewood Cliffs, N.J., 240 pp.

Greenspan, H. F., O'Kon, J. A., Beasley, K. J., and Ward, J. S. (1989). *Guidelines for Failure Investigation.* ASCE, New York, 221 pp.

Griffin, D. C. (1974). "Kentucky's Experience with Dams and Dam Safety." *Safety of Small Dams, Proceedings of the Engineering Foundation Conference,* Henniker, N.H., published by ASCE, New York, pp. 194–207.

"Guajome Ranch House, Vista, California." (1986). *National Historic Landmark Condition Assessment Report,* Preservation Assistance Division, National Park Service, Washington, D.C.

Hammer, M. J., and Thompson, O. B. (1966). "Foundation Clay Shrinkage Caused by Large Trees." *Journal of Soil Mechanics and Foundation Division, ASCE,* vol. 92, no. 6, pp. 1–17.

Hansen, M. J. (1984). "Strategies for Classification of Landslides," in *Slope Instability.* John Wiley & Sons, New York, pp. 1–25.

Hansen, W. R. (1965). *Effects of the Earthquake of March 27, 1964 at Anchorage, Alaska.* Geological Survey Professional Paper 542-A, U.S. Department of the Interior, Washington, D.C.

Harr, E. (1962). *Groundwater and Seepage.* McGraw-Hill, New York, 315 pp.

Holtz, R. D., and Kovacs, W. D. (1981). *An Introduction to Geotechnical Engineering,* Prentice-Hall, Englewood Cliffs, N.J., 733 pp.

Holtz, W. G. (1984). "The Influence of Vegetation on the Swelling and Shrinkage of Clays in the United States of America." *The Influence of Vegetation on Clays,* Thomas Telford, London, pp. 69–73.

Holtz, W. G., and Gibbs, H. J. (1956). "Engineering Properties of Expansive Clays." *Transactions ASCE,* vol. 121, pp. 641–677.

Housner, G. W. (1970). "Strong Ground Motion." Chapter 4 of *Earthquake Engineering,* Robert L. Wiegel, coordinating editor. Prentice-Hall, Englewood Cliffs, N.J., pp. 75–92.

Houston, S. L., and Walsh, K. D. (1993). "Comparison of Rock Correction Methods for Compaction of Clayey Soils." *Journal of Geotechnical Engineering, ASCE,* vol. 119, no. 4, pp. 763–778.

Hurst, W. D. (1968). "Experience in the Winnipeg Area with Sulphate-Resisting Cement Concrete." *Performance of Concrete, Resistance of Concrete to Sulphate and Other Environmental Conditions,* University of Toronto Press, Toronto, pp. 125–134.

Hvorslev, M. J. (1949). *Subsurface Exploration and Sampling of Soils for Civil Engineering Purposes.* Waterways Experiment Station, Vicksburg, Miss., 465 pp.

Ishihara, K. (1993). "Liquefaction and Flow Failure During Earthquakes." *Geotechnique,* vol. 43, no. 3, London, pp. 351–415.

Janbu, N. (1957). "Earth Pressure and Bearing Capacity Calculation by Generalized Procedure of Slices." *Proceedings of the 4th International Conference on Soil Mechanics and Foundation Engineering,* London, vol. 2, pp. 207–212.

Janbu, N. (1968). "Slope Stability Computations." *Soil Mechanics and Foundation Engineering Report,* The Technical University of Norway, Trondheim.

Janney, J. R., Vince, C. R., and Madsen, J. D. (1996). "Claims Analysis from Risk-Retention Professional Liability Group." *Journal of Performance of Constructed Facilities,* ASCE, vol. 10, no. 3, pp. 115–122.

Jennings, J. E. (1953). "The Heaving of Buildings on Desiccated Clay." *Proceedings of the 3d International Conference on Soil Mechanics and Foundation Engineering,* vol. 1, Zurich, pp. 390–396.

Jennings, J. E., and Knight, K. (1957). "The Additional Settlement of Foundations Due to a Collapse of Structure of Sandy Subsoils on Wetting." *Proceedings of the Fourth International Conference on Soil Mechanics and Foundation Engineering,* vol. 1, London, pp. 316–319.

Johnpeer, G. D. (1986). "Land Subsidence Caused by Collapsible Soils in Northern New Mexico." *Ground Failure,* National Research Council, Committee on Ground Failure Hazards, vol. 3, Washington, D.C., 24 pp.

Johnson, A. M., and Hampton, M. A. (1969). "Subaerial and Subaqueous Flow of Slurries." Final Report, U.S. Geological Survey (USGS) Contract no. 14-08-0001-10884, USGS, Boulder, Colo.

Johnson, A. M., and Rodine, J. R. (1984). "Debris Flow." *Slope Instability,* John Wiley & Sons, New York, pp. 257–361.

Johnson, L. D. (1980). "Field Test Sections on Expansive Soil." *Fourth International Conference on Expansive Soils,* Denver, Colo., published by ASCE, pp. 262–283.

Jones, D. E., and Holtz, W. G. (1973). "Expansive Soils—The Hidden Disaster." *Civil Engineering,* vol. 43, November 8.

Jubenville, D. M., and Hepworth, R. (1981). "Drilled Pier Foundation in Shale, Denver, Colorado Area." *Proceedings of the Session on Drilled Piers and Caissons,* ASCE, St. Louis, Mo.

Kaplar, C. W. (1970). "Phenomenon and Mechanism of Frost Heaving." Highway Research Record 304, pp. 1–13.

Kassiff, G., and Baker, R. (1971). "Aging Effects on Swell Potential of Compacted Clay." *Journal of the Soil Mechanics and Foundation Division,* ASCE, vol. 97, no. SM3, pp. 529–540.

Kennedy, M. P. (1975). "Geology of Western San Diego Metropolitan Area, California." Bulletin 200, California Division of Mines and Geology, Sacramento, 39 pp.

Kennedy, M. P., and Tan, S. S. (1977). "Geology of National City, Imperial Beach and Otay Mesa Quadrangles, Southern San Diego Metropolitan Area, California." *Map Sheet 29,* California Division of Mines and Geology, Sacramento, Calif., 1 sheet.

Kerwin, S. T., and Stone, J. J. (1997)."Liquefaction Failure and Remediation: King Harbor Redondo Beach, California." *Journal of Geotechnical and Geoenvironmental Engineering,* ASCE, vol. 123, no. 8, pp. 760–769.

Kisters, F. H., and Kearney, F. W. (1991). "Evaluation of Civil Works Metal Structures." Technical Report REMR-CS-31. Department of the Army, Army Corps of Engineers, Washington, D.C.

Kononova, M. M. (1966). *Soil Organic Matter,* 2d ed., Pergamon Press, Oxford, England.

Kratzsch, H. (1983). *Mining Subsidence Engineering.* Springer-Verlag, Berlin, 543 pp.

Ladd, C. C., Foote, R., Ishihara, K., Schlosser, F., and Poulos, H. G. (1977). "Stress-deformation and Strength Characteristics." *State-of-the-Art Report, Proceedings, Ninth International Conference on Soil Mechanics and Foundation Engineering,* International Society of Soil Mechanics and Foundation Engineering, Tokyo, vol. 2, pp. 421–494.

Ladd, C. C., and Lambe, T. W. (1961). "The Identification and Behavior of Expansive Clays." *Proceedings, Fifth International Conference on Soil Mechanics and Foundation Engineering,* Paris, vol. 1.

Lambe, T. W. (1951). *Soil Testing for Engineers.* John Wiley and Sons, New York, 165 pp.

Lambe, T. W., and Whitman, R. V. (1969). *Soil Mechanics.* John Wiley & Sons, New York, 553 pp.

Lawson, A. C., et al. (1908). *The California Earthquake of April 18, 1906—Report of the State Earthquake Investigation Commission,* vol. 1, part 1, pp. 1–254; part 2, pp. 255–451. Carnegie Institution of Washington, Publication 87.

Lawton, E. C. (1996). "Nongrouting Techniques." *Practical Foundation Engineering Handbook.* Robert W. Brown, ed., McGraw-Hill, New York, Sec. 5, pp. 5.3–5.276.

Lawton, E. C., Fragaszy, R. J., and Hardcastle, J. H. (1989). "Collapse of Compacted Clayey Sand." *Journal of Geotechnical Engineering,* ASCE, vol. 115, no. 9, pp. 1252–1267.

Lawton, E. C., Fragaszy, R. J., and Hardcastle, J. H. (1991). "Stress Ratio Effects on Collapse of Compacted Clayey Sand." *Journal of Geotechnical Engineering,* ASCE, vol. 117, no. 5, pp. 714–730.

Lawton, E. C., Fragaszy, R. J., and Hetherington, M. D. (1992). "Review of Wetting Induced Collapse in Compacted Soil." *Journal of Geotechnical Engineering,* ASCE, vol. 118, no. 9, pp. 1376–1394.

Lea, F. M. (1971). *The Chemistry of Cement and Concrete,* 1st American ed., Chemical Publishing, New York.

Leighton and Associates (1991). "Geotechnical Investigation, Desert View Drive, Ground Motion Study, San Diego, California." Project No. 4910786-01, prepared for City of San Diego, November 18, 12 pp.

Leonards, G. A. (1962). *Foundation Engineering.* McGraw-Hill, New York, 1136 pp.

Leonards, G. A. (1982). "Investigation of Failures." *Journal of the Geotechnical Engineering Division,* ASCE, vol. 99, no. GT2, February.

Lin, G., Bennett, R. M., Drumm, E. C., and Triplett, T. L. (1995). "Response of Residential Test Foundations to Large Ground Movements." *Journal of Performance of Constructed Facilities,* ASCE, vol. 9, no. 4, pp. 319–329.

Lytton, R. L., and Dyke, L. D. (1980). "Creep Damage to Structures on Expansive Clays Slopes." *Fourth International Conference on Expansive Soils,* ASCE, vol. 1, New York, 284–301.

Maksimovic, M. (1989). "Nonlinear Failure Envelope for Soils." *Journal of Geotechnical Engineering,* ASCE, vol. 115, no. 4, pp. 581–586.

Marino, G. G., Mahar, J. W., and Murphy, E. W. (1988). "Advanced Reconstruction for Subsidence-Damaged Homes." *Mine Induced Subsidence: Effects on Engineered Structures,* H. J. Siriwardane, ed. ASCE, New York, pp. 87–106.

Marsh, E. T., and Walsh, R. K. (1996). "Common Causes of Retaining-Wall Distress: Case Study." *Journal of Performance of Constructed Facilities,* ASCE, vol. 10, no. 1, pp. 35–38.

Mather, B. (1968). "Field and Laboratory Studies of the Sulphate Resistance of Concrete." *Performance of Concrete, Resistance of Concrete to Sulphate and Other Environmental Conditions,* University of Toronto Press, Toronto, pp. 66–76.

Matson, J. V. (1994). *Effective Expert Witnessing,* 2d ed. Lewis Publishers, Boca Raton, Fla., 210 pp.

McElroy, C. H. (1987). "The Use of Chemical Additives to Control the Erosive Behavior of Dispersed Clays." *Engineering Aspects of Soil Erosion, Dispersive Clays and Loess,* Geotechnical Special Publication No. 10, C. W. Lovell and R. L. Wiltshire, eds. ASCE, New York, pp. 1–16.

Meehan, R. L., Chun, B., Sang-wuk, J., King, S., Ronold, K., and Yang, F. (1993). "Contemporary Model of Civil Engineering Failures." *Journal of Professional Issues in Engineering Education and Practice,* ASCE, vol. 119, no. 2, pp. 138–146.

Meehan, R. L., and Karp, L. B. (1994). "California Housing Damage Related to Expansive Soils." *Journal of Performance of Constructed Facilities,* ASCE, vol. 8, no. 2, pp. 139–157.

Mehta, P. K. (1976). Discussion of "Combating Sulfate Attack in Corps of Engineers Concrete Construction" by Thomas J. Reading, *ACI Journal Proceedings,* vol. 73, no. 4, pp. 237–238.

Merfield, P. M. (1992). "Surficial Slope Failures: the Role of Vegetation and Other Lessons from Rainstorms." *Engineering Geology Practice in Southern California,* B. W. Pipkin and R. J. Proctor, eds., Star Publishing, Association of Engineering Geologists, Southern California Section, Special Publication No. 4, pp. 613–627.

Middlebrooks, T. A. (1953). "Earth Dam Practice in the United States." *Transactions, American Society of Civil Engineers,* centennial volume, 697 pp.

Miller, T. E. (1993). *California Construction Defect Litigation, Residential and Commercial,* 2d ed. John Wiley & Sons, New York, 735 pp.

Mitchell, J. K. (1970). "In-Place Treatment of Foundation Soils," *Journal of the Soil Mechanics and Foundation Division,* ASCE, vol. 96, no. SM1, pp. 73–110.

Monahan, E. J. (1986). *Construction of and on Compacted Fills.* John Wiley & Sons, New York, 200 pp.

Mottana, A., Crespi, R., and Liborio, G. (1978). *Rocks and Minerals.* Simon & Schuster, New York, 607 pp.

Nadjer, J., and Werno, M. (1973). "Protection of Buildings on Expansive Clays." *Proceedings of the Third International Conference on Expansive Soils,* Haifa, Israel, vol. 1, pp. 325–334.

Narver (1993). "A/E Risk Review," *Professional Liability Agents Network,* vol. 3, no. 7.

National Coal Board (1975). *Subsidence Engineers Handbook.* National Coal Board Mining Department, National Coal Board, London, 111 pp.

National Research Council (1985). *Reducing Losses from Landsliding in the United States.* Committee on Ground Failure Hazards, Commission on Engineering and Technical Systems, National Academy Press, Washington, D.C., 41 pp.

NAVFAC DM-7.1 (1982). *Soil Mechanics.* Design Manual 7.1, Department of the Navy, Naval Facilities Engineering Command, Alexandria, Va., 364 pp.

NAVFAC DM-7.2 (1982). *Foundations and Earth Structures.* Design Manual 7.2, Department of the Navy, Naval Facilities Engineering Command, Alexandria, Va., 253 pp.

NAVFAC DM-21.3 (1978). *Flexible Pavement Design for Airfields,* Design Manual 21.3, Department of the Navy, Naval Facilities Engineering Command, Alexandria, Va., 98 pp.

Neary, D. G., and Swift, L. W. (1987). "Rainfall Thresholds for Triggering a Debris Avalanching Event in the Southern Appalachian Mountains." *Debris Flows/Avalanches: Process, Recognition, and Mitigation,* The Geological Society of America, Boulder, Colo. pp. 81–92.

Nelson, J. D., and Miller, D. J. (1992). *Expansive Soils, Problems and Practice in Foundation and Pavement Engineering.* John Wiley & Sons, New York, 259 pp.

Noon, R. (1992). *Introduction to Forensic Engineering.* CRC Press, Boca Raton, Florida, 205 pp.

Norris, R. M., and Webb, R. W. (1990). *Geology of California,* 2d ed. John Wiley & Sons, New York, 541 pp.

NSF (1992). "Quantitative Nondestructive Evaluation for Constructed Facilities." *Announcement Fiscal Year 1992.* National Science Foundation, Directorate for Engineering, Division of Mechanical and Structural Systems, Washington, D.C.

Oldham, R. D. (1899). "Report on the Great Earthquake of 12th June, 1897." India Geologic Survey Memorial, Publication 29, 379 pp.

Oliver, A. C. (1988). *Dampness in Buildings.* Internal and Surface Waterproofers, Nichols Publishing, New York, 221 pp.

Ortigao, J. A. R., Loures, T. R. R., Nogueiro, C., and Alves, L. S. (1997). "Slope Failures in Tertiary Expansive OC Clays." *Journal of Geotechnical and Geoenvironmental Engineering,* ASCE, vol. 123, no. 9, pp. 812–817.

Owens, D. T. (1993). "Red Line Cost Overruns." *Los Angeles Times,* editorial section, November 20, 1993, Los Angeles.

Patton, J. H. (1992). "The Nuts and Bolts of Litigation." *Engineering Geology Practice in Southern California,* B. W. Pipkin and R. J. Proctor, eds., Star Publishing, Belmont, Calif., pp. 339–359.

Peck, R. B., Hanson, W. E., and Thornburn, T. H. (1974). *Foundation Engineering,* 2d ed. John Wiley & Sons, New York, 514 pp.

Peckover, F. L. (1975). "Treatment of Rock Falls on Railway Lines." *American Railway Engineering Association,* Bulletin 653, Chicago, pp. 471–503.

Peng, S. S. (1986). *Coal Mine Ground Control.* 2d ed. John Wiley & Sons, New York, 491 pp.

Peng, S. S. (1992). *Surface Subsidence Engineering.* Society for Mining, Metallurgy and Exploration, Littleton, Colo.

Perry, D., and Merschel, S. (1987). "The Greening of Urban Civilization." *Smithsonian,* vol. 17, no. 10, pp. 72–79.

Perry, E. B. (1987). "Dispersive Clay Erosion at Grenada Dam, Mississippi." *Engineering Aspects of Soil Erosion, Dispersive Clays and Loess,* Geotechnical Special Publication No. 10, C. W. Lovell and R. L. Wiltshire, eds. ASCE, New York, pp. 30–45.

Petersen, E. V. (1963). "Cave-in!" *Roads and Engineering Construction,* November, pp. 25–33.

Piteau, D. R., and Peckover, F. L. (1978). "Engineering of Rock Slopes." *Landslides, Analysis and Control,* Special Report 176, Transportation Research Board, National Academy of Sciences, chap. 9, pp. 192–228.

Poh, T. Y., Wong, I. H., and Chandrasekaran, B. (1997). "Performance of Two Propped Diaphragm Walls in Stiff Residual Soils." *Journal of Performance of Constructed Facilities,* ASCE, vol. 11, no. 4, pp. 190–199.

Post-Tensioning Institute (1996). "Design and Construction of Post-tensioned Slabs-on-Ground," 2d ed. Report, Phoenix, Ariz., 101 pp.

Pradel, D., and Raad, G. (1993). "Effect of Permeability of Surficial Stability of Homogeneous Slopes." *Journal of Geotechnical Engineering,* ASCE, vol. 119, no. 2, pp. 315–332.

Purkey, B. W., Duebendorfer, E. M., Smith, E. I., Price, J. G., Castor, S. B. (1994). *Geologic Tours in the Las Vegas Area,* Nevada Bureau of Mines and Geology, Special Publication 16, Las Vegas, 156 pp.

Rathje, W. L., and Psihoyos, L. (1991). "Once and Future Landfills." *National Geographic,* vol. 179, no. 5, pp. 116–134.

Ravina, I. (1984). "The Influence of Vegetation on Moisture and Volume Changes." *The Influence of Vegetation on Clays,* Thomas Telford, London, pp. 62–68.

Reading, T. J. (1975). "Combating Sulfate Attack in Corps of Engineering Concrete Construction." *Durability of Concrete,* SP47, American Concrete Institute, Detroit, pp. 343–366.

Reed, M. A., Lovell, C. W., Altschaeffl, A. G., and Wood, L. E. (1979). "Frost Heaving Rate Predicted from Pore Size Distribution," *Canadian Geotechnical Journal,* vol. 16, no. 3, pp. 463–472.

Reese, L. C., Owens, M., and Hoy, H. (1981). "Effects of Construction Methods on Drilled Shafts." *Drilled Piers and Caissons,* M. W. O'Neill, ed. ASCE, New York, pp. 1–18.

Reese, L. C., and Tucker, K. L. (1985). "Bentonite Slurry in Concrete Piers." *Drilled Piers and Caisson II,* C. N. Baker, ed. ASCE, New York, pp. 1–15.

Rens, K. L., Wipf, T. J., and Klaiber, F. W. (1997). "Review of Nondestructive Evaluation Techniques of Civil Infrastructure." *Journal of Performance of Constructed Facilities,* ASCE, vol. 11, no. 4, pp. 152–160.

Rice, R. J. (1988). *Fundamentals of Geomorphology,* 2d ed., John Wiley and Sons, New York.

Ritchie, A. M. (1963). "Evaluation of Rockfall and Its Control." *Highway Research Record 17,* Highway Research Board, Washington, D.C., pp. 13–28.

Rogers, J. D. (1992). "Recent Developments in Landslide Mitigation Techniques." Chapter 10 of *Landslides/Landslide Mitigation,* J. E. Slosson, G. G. Keene, and J. A. Johnson, eds. The Geological Society of America, Boulder, Colo., pp. 95–118.

Rollins, K. M., Rollins, R. C., Smith, T. D., and Beckwith, G. H. (1994). "Identification and Characterization of Collapsible Gravels." *Journal of Geotechnical Engineering,* ASCE, vol. 120, no. 3, pp. 528–542.

Roofing Equipment, Inc. (undated). *Moisture Test Kit Pamphlet.* Roofing Equipment, Inc., 4 pp.

Ross, C. S., and Smith, R. L. (1961). *Ash-Flow Tuffs, Their Origin, Geologic Relations and Identification,* U.S. Geological Survey Professional Paper 366, U. S. Geological Survey, Denver, Colo.

Rutledge, P. C. (1944). "Relation of Undisturbed Sampling to Laboratory Testing." *Transactions,* ASCE, vol. 109, pp. 1162–1163.

Sanglerat, G. (1972). *The Penetrometer and Soil Exploration.* Elsevier Scientific, New York, 464 pp.

Savage, J. C., and Hastie, L. M. (1966). "Surface Deformation Associated with Dip-Slip Faulting." *Journal of Geophysical Research,* vol. 71, no. 20, pp. 4897–4904.

Saxena, S. K., Lourie, D. E., and Rao, J. S. (1984). "Compaction Criteria for Eastern Coal Waste Embankments." *Journal of Geotechnical Engineering,* ASCE, vol. 110, no. 2, pp. 262–284.

Schlager, N. (1994). *When Technology Fails.* "St. Francis Dam Failure." Gale Research, Detroit, pp. 426–430.

Schnitzer, M., and Khan, S. U. (1972). *Humic Substances in the Environment.* Marcel Dekker, New York.

Schuster, R. L. (1986). *Landslide Dams: Processes, Risk, and Mitigation,* Geotechnical Special Publication No. 3. *Proceedings of Geotechnical Session,* Seattle. Published by ASCE, New York, 164 pp.

Schuster, R. L., and Costa, J. E. (1986). "A Perspective on Landslide Dams." *Landslide Dams: Processes, Risk, and Mitigation,* Geotechnical Special Publication No. 3. *Proceedings of Geotechnical Session,* Seattle. Published by ASCE, New York, pp. 1–20.

Schutz, R. J. (1984). "Properties and Specifications for Epoxies Used in Concrete Repair." *Concrete Construction Magazine,* Concrete Construction Publications, Addison, Ill. pp. 873–878.

Seed, H. B. (1970). "Soil Problems and Soil Behavior." Chapter 10 of *Earthquake Engineering,* Robert L. Wiegel, coordinating editor. Prentice-Hall, Englewood Cliffs, N.J., pp. 227–252.

Seed, H. B., Woodward, R. J., and Lundgren, R. (1962). "Prediction of Swelling Potential for Compacted Clays." *Journal of Soil Mechanics and Foundations Division,* ASCE, vol. 88, no. SM3, pp. 53–87.

Shannon and Wilson, Inc. (1964). *Report on Anchorage Area Soil Studies, Alaska, to U.S. Army Engineer District, Anchorage, Alaska.* Seattle.

Sherard, J. L. (1972). "Study of Piping Failures and Eroding Damage from Rain in Clay Dams in Oklahoma and Mississippi." U.S. Department of Agriculture, Soil Conservation Service, Washington, D.C.

Sherard, J. L., Decker, R. S., and Ryker, N. L. (1972). "Piping in Earth Dams of Dispersive Clay." *Proceedings of the Specialty Conference on Performance of Earth and Earth-Supported Structures,* vol. 1, part 1, cosponsored by ASCE and Purdue University, Lafayette, Ind., pp. 589–626.

Sherard, J. L., Woodward, R. J., Gizienski, S. F., and Clevenger, W. A. (1963). *Earth and Earth-Rock Dams.* John Wiley and Sons, New York, 725 pp.

Shuirman, G., and Slosson, J. E. (1992). *Forensic Engineering—Environmental Case Histories for Civil Engineers and Geologists.* Academic Press, New York, 296 pp.

Skempton, A. W., and MacDonald, D. H. (1956). "The Allowable Settlement of Buildings." *Proceedings of the Institution of Civil Engineers,* Part III. The Institution of Civil Engineers, London, no. 5, pp. 727–768.

Slope Indicator (1996). *Geotechnical and Structural Instrumentation,* Slope Indicator Co., Bothell, Wash., 92 pp.

Slough-in Incident Report. (1986). Shoring Design Engineers. November 24.

Smith, D. D., and Wischmeier, W. H. (1957). "Factors Affecting Sheet and Rill Erosion." *Transactions of the American Geophysical Union,* vol. 38, no. 6, pp. 889–896.

Smith, R. L. (1960). *Zones and Zonal Variations in Welded Ash Flows,* U.S. Geological Survey Professional Paper 354-F, U.S. Geological Survey, Denver.

Snethen, D. R. (1979). *Technical Guidelines for Expansive Soils in Highway Subgrades.* U.S. Army Engineering Waterway Experiment Station, Vicksburg, Miss., Report No. FHWA-RD-79-51.

Sowers, G. B., and Sowers, G. F. (1970). *Introductory Soil Mechanics and Foundations.* 3d ed. Macmillan, New York, 556 pp.

Sowers, G. F. (1962). "Shallow Foundations," Chap. 6 from *Foundation Engineering,* G. A. Leonards, ed. McGraw-Hill, New York.

Sowers, G. F. (1974). "Dam Safety Legislation: A Solution or a Problem." *Safety of Small Dams, Proceedings of the Engineering Foundation Conference,* Henniker, N.H. Published by ASCE, New York, pp. 65–100.

Sowers, G. F. (1979). *Soil Mechanics and Foundations: Geotechnical Engineering,* 4th ed., Macmillan, New York.

Sowers, G. F. (1997). *Building on Sinkholes: Design and Construction of Foundations in Karst Terrain,* ASCE Press, New York.

Sowers, G. F., and Royster, D. L. (1978). "Field Investigation." Chapter 4 of *Landslides, Analysis and Control,* Special Report 176, R. L. Schuster and R. J. Krizek, eds. Transportation Research Board, National Academy of Sciences, Washington, D.C., pp. 81–111.

Spencer, E. W. (1972). *The Dynamics of the Earth, An Introduction to Physical Geology.* Thomas Y. Crowell, New York, 649 pp.

Standard Specifications for Public Works Construction (1997). Bni Building News, Los Angeles, commonly known as the "Green Book," 761 pp.

Standards Presented to California Occupational Safety and Health Standard Board, Sections 1504 and 1539–1547 (1991). California Occupational Safety and Health Standard Board, San Francisco, July.

Stapledon, D. H., and Casinader, R. J. (1977). "Dispersive Soils at Sugarloaf Dam Site Near Melbourne, Australia." *Dispersive Clays, Related Piping, and Erosion in Geotechnical Projects.* STP 623, American Society for Testing and Materials, Philadelphia, pp. 432–466.

Stark, T. D., and Eid, H. T. (1994). "Drained Residual Strength of Cohesive Soils." *Journal of Geotechnical Engineering,* ASCE, vol. 120, no. 5, pp. 856–871.

Steinbrugge, K. V. (1970). "Earthquake Damage and Structural Performance in the United States." Chapter 9 of *Earthquake Engineering,* Robert L. Wiegel, coordinating editor. Prentice-Hall, Englewood Cliffs, N.J., pp. 167–226.

Stokes, W. C., and Varnes, D. J. (1955). *Glossary of Selected Geologic Terms.* Colorado Scientific Society Proceedings, vol. 116, Denver, 165 pp.

Sweet, J. (1970). *Legal Aspects of Architecture, Engineering and Construction Process.* West Publishing, St. Paul, Minn., 953 pp.

Tadepalli, R., and Fredlund, D. G. (1991). "The Collapse Behavior of Compacted Soil During Inundation." *Canadian Geotechnical Journal,* vol. 28, no. 4, pp. 477–488.

Terzaghi, K. (1938). "Settlement of Structures in Europe and Methods of Observation," *Transactions ASCE,* vol. 103, p. 1432.

Terzaghi, K., and Peck, R. B. (1967). *Soil Mechanics in Engineering Practice,* 2d ed. John Wiley and Sons, New York, 729 pp.

Thompson, L. J., and Tanenbaum, R. J. (1977). "Survey of Construction Related Trench Cave-Ins." *Journal of the Construction Division,* ASCE, vol. 103, no. CO3, September.

Transportation Research Board (1977). *Rapid-Setting Materials for Patching Concrete.* National Cooperative Highway Research Program Synthesis of Highway Practice 45, published by National Academy of Sciences, Washington, D.C.

"Trench and Excavation Safety Guide" (1984). Publication S-358, California Department of Industrial Relations/California Occupational Safety and Health Administration, San Francisco.

Tucker, R. L., and Poor, A. R, (1978). "Field Study of Moisture Effects on Slab Movements." *Journal of Geotechnical Engineering Division,* ASCE, vol. 104, no. 4, pp. 403–414.

Turnbull, W. J., and Foster, C. R. (1956). "Stabilization of Materials by Compaction." *Journal of Soil Mechanics and Foundation Division,* ASCE, vol. 82, no. 2, pp. 934.1–934.23.

Tuthill, L. H. (1966). "Resistance to Chemical Attack-Hardened Concrete." *Significance of Tests and Properties of Concrete and Concrete-Making Materials,* STP-169A, ASTM, Philadelphia, pp. 275–289.

Uniform Building Code (1997). International Conference of Building Officials, 3 volumes, Whittier, Calif.

Van der Merwe, C. P., and Ahronovitz, M. (1973). "The Behavior of Flexible Pavements on Expansive Soils." *Third International Conference on Expansive Soil*, Haifa, Israel.

Varnes, D. J. (1978). "Slope Movement and Types and Processes." *Landslides: Analysis and Control*, Transportation Research Board, National Academy of Sciences, Washington, D.C., Special Report 176, chap. 2, pp. 11–33.

WFCA (1984). "Moisture Guidelines for the Floor Covering Industry." *WFCA Management Guidelines*, Western Floor Covering Association, Los Angeles.

Wahls, H. E. (1994). "Tolerable Deformations." *Vertical and Horizontal Deformations of Foundations and Embankments*, Geotechnical Special Publication No. 40, ASCE, New York, pp. 1611–1628.

Waldron, L. J. (1977). "The Shear Resistance of Root-Permeated Homogeneous and Stratified Soil." *Soil Science Society of America*, vol. 41, no. 5, pp. 843–849.

Wallace, T. (1981). "Preparation of Reports, Field Notes, and Documentation." *Proceedings, Geotechnical Construction Loss Prevention Seminar*, Santa Clara, Calif.

Warner, J. (1982). "Compaction Grouting—The First Thirty Years." *Proceedings of the Conference on Grouting in Geotechnical Engineering*, W. H. Baker, ed. ASCE, New York, pp. 694–707.

Warriner, J. E. (1957). *English Grammar and Composition*. Harcourt, Brace, New York, 692 pp.

Watry, S. M., and Ehlig, P. L. (1995). "Effect of Test Method and Procedure on Measurements of Residual Shear Strength of Bentonite from the Portuguese Bend Landslide." *Clay and Shale Slope Instability*, W. C. Haneberg and S. A. Anderson, eds., Geological Society of America, Reviews in Engineering Geology, vol. 10, Boulder, Colo., pp. 13–38.

Whitlock, A. R., and Moosa, S. S. (1996). "Foundation Design Considerations for Construction on Marshlands." *Journal of Performance of Constructed Facilities*, ASCE, vol. 10, no. 1, pp. 15–22.

Williams, A. A. B. (1965). "The Deformation of Roads Resulting from Moisture Changes in Expansive Soils in South Africa." *Symposium Proceedings, Moisture Equilibria and Moisture Changes in Soils Beneath Covered Areas*, G. D. Aitchison, ed. Butterworths, Australia, pp. 143–155.

Winterkorn, H. F., and Fang, H. (1975). *Foundation Engineering Handbook*. Van Nostrand Reinhold, New York, 751 pp.

Woodward, R. J., Gardner, W. S., and Greer, D. M. (1972). "Design Considerations," *Drilled Pier Foundations*, D. M. Greer, ed. McGraw-Hill, New York, pp. 50–52.

Wu, T. H., Randolph, B. W., and Huang, C. (1993). "Stability of Shale Embankments." *Journal of Geotechnical Engineering*, ASCE, vol. 119, no. 1, pp. 127–146.

Yegian, M. K., Ghahraman, V. G., and Gazetas, G. (1994). "1988 Armenia Earthquake. II: Damage Statistics versus Geologic and Soil Profiles." *Journal of Geotechnical Engineering*, ASCE, vol. 120, no. 1, pp. 21–45.

Yong, R. N., and Warkentin, B. P. (1975). *Soil Properties and Behavior*, Elsevier, New York, 449 pp.

Zaruba, Q., and Mencl, V. (1969). *Landslides and Their Control*. Elsevier, New York, 205 pp.

Glossary 1

Engineering Geology,

and Subsurface

Exploration Terminology

Introduction

The following is a list of commonly used geotechnical engineering and engineering geology terms and definitions. The glossary has been divided into four main categories:

1. Engineering Geology and Subsurface Exploration Terminology
2. Laboratory Testing Terminology
3. Terminology for Engineering Analysis and Computations
4. Construction and Grading Terminology

References

References used for the Glossary include:

- *Glossary of Selected Geologic Terms with Special Reference to Their Use in Engineering* (1955)
- *Soil Mechanics in Engineering Practice* (1967)
- *Soil Mechanics* (1969)
- *Essentials of Soil Mechanics and Foundations* (1977)
- *An Introduction to Geotechnical Engineering* (1981)
- NAVFAC DM-7.1 (1982), NAVFAC DM-7.2 (1982), and NAVFAC DM-7.3 (1983)
- *Thickness Design—Asphalt Pavements for Highways and Streets* (1984)

- *Orange County Grading and Excavation Code* (1993)
- *Foundation Design, Principles and Practice* (1994)
- *Standard Specifications for Highway Bridges* (1996)
- *Caterpillar Performance Handbook* (1997)
- *Uniform Building Code* (1997)
- American Society for Testing and Materials (ASTM D 653-97 and ASTM D 4439-95, 1998)

Basic terms

Civil Engineer A professional engineer who is registered to practice in the field of civil works.

Civil Engineering The application of the knowledge of the forces of nature, principles of mechanics, and properties of materials for the evaluation, design, and construction of civil works for the beneficial uses of mankind.

Engineering Geologist A geologist who is experienced and knowledgeable in the field of engineering geology.

Engineering Geology The application of geologic knowledge and principles in the investigation and evaluation of naturally occurring rock and soil for use in the design of civil works.

Geologist An individual educated and trained in the field of geology.

Geotechnical Engineer An licensed individual who performs an engineering evaluation of earth materials including soil, rock, groundwater, and man-made materials and their interaction with earth retention systems, structural foundations, and other civil engineering works.

Geotechnical Engineering A subdiscipline of civil engineering. Geotechnical engineering requires a knowledge of engineering laws, formulas, construction techniques, and the performance of civil engineering works influenced by earth materials. Geotechnical engineering encompasses many of the engineering aspects of soil mechanics, rock mechanics, foundation engineering, geology, geophysics, hydrology, and related sciences.

Rock Mechanics The application of the knowledge of the mechanical behavior of rock to engineering problems dealing with rock. Rock mechanics overlaps with engineering geology.

Soil Mechanics The application of the laws and principles of mechanics and hydraulics to engineering problems dealing with soil as an engineering material.

Soils Engineer Synonymous with geotechnical engineer (*see* Geotechnical Engineer).

Soils Engineering Synonymous with geotechnical engineering (*see* Geotechnical Engineering).

Engineering Geology and Subsurface Exploration Terminology

Adobe   Sun-dried bricks composed of mud and straw. Abode is commonly used for construction in the southwestern United States and in Mexico.

Aeolian (or Eolian)   Particles of soil that have been deposited by the wind. Aeolian deposits include dune sands and loess.

Alluvium   Detrital deposits resulting from the flow of water, including sediments deposited in river beds, canyons, flood plains, lakes, fans at the foot of slopes, and estuaries.

Aquiclude   A relatively impervious rock or soil strata that will not transmit groundwater fast enough to furnish an appreciable supply of water to a well or spring.

Aquifer   A relatively pervious rock or soil stratum that will transmit groundwater fast enough to furnish an appreciable supply of water to a well or spring.

Artesian water   Groundwater that is under pressure and is confined by impervious material. If the trapped pressurized water is released, such as by drilling a well, the water will rise above the groundwater table and may even rise above the ground surface.

Ash   Fine fragments of rock, between 4 mm and 0.25 mm in size, that originated as airborne debris from explosive volcanic eruptions.

Badlands   An area, large or small, characterized by extremely intricate and sharp erosional sculpture. Badlands occur chiefly in arid or semiarid climates where the rainfall is concentrated in sudden heavy showers. They may, however, occur in humid regions where vegetation has been destroyed, or where soil and coarse detritus are lacking.

Bedding   The arrangement of rock in layers, strata, or beds.

Bedrock   A more or less solid, relatively undisturbed rock in place either at the surface or beneath deposits of soil.

Bentonite   A soil or formational material that has a high concentration of the clay mineral montmorillonite. Bentonite is usually characterized by high swelling upon wetting. The term *bentonite* also refers to manufactured products that have a high concentration of montmorillonite, for example, bentonite pellets.

Bit   A device that is attached to the end of the drill stem and is used as a cutting tool to bore into soil and rock.

Bog   A peat-covered area with a high groundwater table. The surface is often covered with moss, and it tends to be nutrient-poor and acidic.

Boring   A method of investigating subsurface conditions by drilling a hole into the earth materials. Usually soil and rock samples are extracted from the boring. Field tests, such as the standard penetration test (SPT) and the vane shear test (VST), can also be performed in the boring.

Boring Log A written record of the materials penetrated during the subsurface exploration. See Figs. 3.16 and 6.48 for other types of information that should be recorded on the boring log.

Boulder A large detached rock fragment with an average dimension greater than 300 mm (12 in.).

California Bearing Ratio (CBR) The CBR can be determined for soil in the field or soil compacted in the laboratory. See ASTM for specific details on the procedure for determining the CBR. The CBR is frequently used for the design of roads and airfields.

Casing A steel pipe that is temporarily inserted into a boring or drilled shaft in order to prevent the adjacent soil from caving.

Cobble A rock fragment, usually rounded or semirounded, with an average dimension between 75 and 305 mm (3 and 12 in.).

Cohesionless Soil A soil, such as a clean gravel or sand, that when unconfined falls apart in either a wet or dry state.

Cohesive Soil A soil, such as a silt or clay, that when unconfined has considerable shear strength when dried, and will not fall apart in a saturated state. Cohesive soil is also known as a *plastic soil,* or a soil that has a plasticity index.

Colluvium Generally loose deposits usually found near the base of slopes and brought there chiefly by gravity through slow continuous downhill creep.

Cone Penetration Test (CPT) A field test used to identify and determine the *in situ* properties of soil deposits and soft rock.

   Electric Cone A cone penetrometer that uses electric force transducers built into the apparatus for measuring cone resistance and friction resistance.

   Mechanical Cone A cone penetrometer that uses a set of inner rods to operate a telescoping penetrometer tip and to transmit the resistance force to the surface for measurement.

   Mechanical-Friction Cone A cone penetrometer with the additional capability of measuring the local side friction component of penetration resistance.

   Piezocone A cone penetrometer with the additional capability of measuring pore water pressure generated during the penetration of the cone.

Core Drilling Also known as *diamond drilling,* the process of cutting out cylindrical rock samples in the field.

Core Recovery (RQD) The RQD is computed by summing the lengths of all pieces of the rock core (NX size) equal to or longer than 10 cm (4 in.) and dividing by the total length of the core run. The RQD is multiplied by 100 to express it as a percentage.

Deposition The geologic process of laying down or accumulating natural material into beds, veins, or irregular masses. Deposition includes mechanical settling (such as sedimentation in lakes), precipitation (such as the evaporation of surface water to form halite), and the accumulation of dead plants and animals (such as in a bog).

Detritus   Any material worn or broken down from rocks by mechanical means.

Diatomaceous Earth   Diatomaceous earth usually consists of fine, white, siliceous powder, composed mainly of diatoms or their remains.

Erosion   The wearing away of the ground surface as a result of the movement of wind, water, and/or ice.

Fault   A fracture in the earth's crust along which movement has occurred. A fault is considered active if movement has occurred within the last 11,000 years (Holocene geologic time).

Fines   The silt- and clay-size particles in the soil.

Fold   Bending or flexure of a layer or layers or rock. Examples of folded rock include anticlines and synclines. Usually folds are created by the massive compression of rock layers.

Fracture   A visible break in a rock mass. Examples includes joints, faults, and fissures.

Geophysical Techniques   Various methods of determining subsurface soil and rock conditions without performing subsurface exploration. A common geophysical technique is to induce a shock wave into the earth and then measure the seismic velocity of the wave's travel through the earth material. The seismic velocity has been correlated with the rippability of the earth material.

Groundwater Table (Phreatic Surface)   The top surface of underground water, the location of which is often determined from piezometers, such as an open standpipe. A perched groundwater table refers to groundwater occurring in an upper zone separated from the main body of groundwater by underlying unsaturated rock or soil.

Horizon   One of the layers of a soil profile that can be distinguished by its texture, color, and structure.

    A Horizon   The uppermost layer of a soil profile which often contains remnants of organic life. Inorganic colloids and soluble materials are often leached from this horizon.

    B Horizon   The layer of a soil profile in which material leached from the overlying A horizon is accumulated.

    C Horizon   Undisturbed parent material from which the overlying soil profile has been developed.

Inclinometer   The horizontal movement preceding or during the movement of slopes can be investigated by successive surveys of the shape and position of flexible vertical casings installed in the ground. The surveys are performed by lowering an inclinometer probe into the flexible vertical casing.

Iowa Borehole Shear Test (BST)   A field test where the device is lowered into an uncased borehole and then expanded against the side walls. The force required to pull the device toward ground surface is measured and, much as in a direct shear test, the shear strength properties of the *in situ* soil can then be determined.

In Situ Used in reference to the original in-place (or *in situ*) condition of the soil or rock.

Karst Topography A type of landform developed in a region of easily soluble limestone. It is characterized by vast numbers of depressions of all sizes; sometimes by great outcrops of limestone ledges, sinks, and other solution passages; an almost total lack of surface streams; and large springs in the deeper valleys.

Kelly A heavy tube or pipe, usually square or rectangular in cross section, that is used to provide a downward load in excavating an auger borehole.

Landslide Mass movement of soil or rock that involves shear displacement along one or several rupture surfaces, which are either visible or may be reasonably inferred.

Landslide Debris Material, generally porous and of low density, produced from instability of natural or man-made slopes.

Leaching The removal of soluble materials in soil or rock by percolating or moving groundwater.

Loess A wind-deposited silt, often having a high porosity and low density, which is often susceptible to collapse of its soil structure upon wetting.

Mineral An inorganic substance that has a definite chemical composition and distinctive physical properties. Most minerals are crystalline solids.

Overburden The soil that overlies bedrock. In other cases, it refers to all material overlying a point of interest in the ground, such as that causing the overburden pressure on a clay layer.

Peat A naturally occurring highly organic deposit derived primarily from plant materials.

Penetration Resistance *See* Standard Penetration Test.

Percussion Drilling A drilling process in which a borehole is advanced by a series of impacts to the drill rods and attached bit.

Permafrost Perennially frozen soil. Also defined as ground that remains below freezing temperatures for 2 or more years. The bottom of permafrost lies at depths ranging from a few feet to over a thousand feet. The *active layer* is defined as the upper few inches to several feet of ground that is frozen in winter but thawed in summer.

Piezometer A device installed for measuring the pore water pressure (or pressure head) at a specific point within the soil mass.

Pit (Test Pit) An excavation made for the purpose of observing subsurface conditions and obtaining soil samples. A pit also refers to an excavation in the surface of the earth from which ore is extracted, such as an open pit mine.

Pressuremeter Test (PST) A field test that involves the expansion of a cylindrical probe within an uncased borehole.

Refusal During subsurface exploration, refusal means an inability to excavate any deeper with the boring equipment. Refusal could be due to many different factors, such as hard rock, boulders, or a layer of cobbles.

Residual Soil Soil derived by in-place weathering of the underlying material.

Rock Rock is a relatively solid mass that has permanent and strong bonds between the minerals. Rock can be classified as being sedimentary, igneous, or metamorphic.

Rotary Drilling A drilling process in which a borehole is advanced by rotation of a drill bit under constant pressure without impact.

Rubble Rough stones of irregular shape and size that are naturally or artificially broken from larger masses of rock. Rubble is often created during quarrying, stone cutting, and blasting.

Screw Plate Compressometer (SPC) A field test in which a plate is screwed down to the desired depth, and, as pressure is applied, the settlement of the plate is measured.

Seep A small area where water oozes from the soil or rock.

Slaking The crumbling and disintegration of earth materials when exposed to air or moisture. Slaking can also refer to the breaking up of dried clay when submerged in water, due either to compression of entrapped air by inwardly migrating water or to the progressive swelling and sloughing off of the outer layers.

Slickensides Surfaces within a soil mass which have been smoothed and striated by shear movements on these surfaces.

Slope Wash Soil and/or rock material that has been transported down a slope by mass wasting assisted by runoff water not confined by channels (*see* Colluvium).

Soil Sediments or other accumulations of mineral particles produced by the physical and chemical disintegration of rocks. Inorganic soil does not contain organic matter, while organic soil contains organic matter.

Soil Profile Developed from subsurface exploration, a cross section of the ground that shows the soil and rock layers. A summary of field and laboratory tests could also be added to the soil profile.

Soil Sampler A device used to obtain soil samples during subsurface exploration. Depending on the inside clearance ratio and the area ratio, soil samples can be either disturbed or undisturbed.

Standard Penetration Test (SPT) A field test that consists of driving a thick-walled sampler (I.D. = 1.5 in., O.D. = 2 in.) into the soil by using a 140-lb hammer falling 30 in. The number of blows to drive the sampler 18 in. is recorded. The N value (penetration resistance) is defined as the number of blows required to drive the sampler from a depth interval of 6 to 18 in.

Strike and Dip Strip and dip refer to a planar structure, such as a shear surface, fault, bed, etc. The strike is the compass direction of a level line drawn on the planar structure. The dip angle is measured between the planar structure and a horizontal surface.

Subgrade Modulus (Modulus of Subgrade Reaction) This value is often obtained from field plate load tests and is used in the design of pavements and airfields.

Test Pit *See* Pit.

Till Material created directly by glaciers, without transportation or sorting by water. Till often consists of particles in a wide range of sizes, including boulders, gravel, sand, and clay.

Topsoil The fertile upper zone of soil which contains organic matter and is usually darker in color and loose.

Vane Shear Test (VST) An *in situ* field test that consists of inserting a four-bladed vane into the borehole and then pushing the vane into the clay deposit located at the bottom of the borehole. Once inserted into the clay, the maximum torque required to rotate the vane and shear the clay is measured. Based on the dimensions of the vane and the maximum torque, the undrained shear strength s_u of the clay can be calculated.

Varved Silt or Varved Clay A lake deposit with alternating thin layers of sand and silt (varved silt) or sand and clay (varved clay). It is formed by the process of sedimentation from the summer to winter months. The sand is deposited during the summer and the silt or clay is deposited in the winter when the lake surface is covered with ice and the water is tranquil.

Wetland Land which has a groundwater table at or near the ground surface, or land that is periodically under water, and supports various types of vegetation that are adapted to a wet environment.

Glossary 2

Laboratory Testing Terminology

Absorption   Defined as the mass of water in the aggregate divided by the dry mass of the aggregate. Absorption is used in soil mechanics for the study of oversize particles or in concrete mix design.

Activity of Clay   The ratio of plasticity index to percent dry mass of the total sample that is smaller than 0.002 mm in grain size. This property is correlated with the type of clay mineral.

Atterberg Limits   Water contents corresponding to different behavior conditions of plastic soil.

Liquid Limit   The water content corresponding to the behavior change between the liquid and plastic state of a soil. The liquid limit is arbitrarily defined as the water content at which a pat of soil, cut by a groove of standard dimensions, will flow together for a distance of 12.7 mm (0.5 in.) under the impact of 25 blows in a standard liquid limit device.

Plastic Limit   The water content corresponding to the behavior change between the plastic and semisolid state of a soil. The plastic limit is arbitrarily defined as the water content at which the soil will just begin to crumble when rolled into a tread approximately 3.2 mm ($\frac{1}{8}$ in.) in diameter.

Shrinkage Limit   The water content corresponding to the behavior change between the semisolid to solid state of a soil. The shrinkage limit is also defined as the water content at which any further reduction in water content will not result in a decrease in volume of the soil mass.

Capillarity   Also known as *capillary action,* the rise of water through a soil due to the fluid property known as *surface tension.* Because of capillarity, the pore water pressures are less than atmospheric values produced by surface tension of pore water acting on the meniscus formed in void spaces between the soil particles. The height of capillary rise is related to the pore size of the soil.

Clay-Size Particles   Clay-size particles are finer than 0.002 mm. Most clay particles are flat or platelike in shape, and as such they have a large surface area. The most common clay minerals belong to the kaolin, montmorillonite, or illite groups.

Coarse-Grained Soil   According to the Unified Soil Classification System, coarse-grained soils have more than 50 percent soil particles (by dry mass) retained on the no. 200 U.S. standard sieve.

Coefficient of Consolidation   A coefficient used in the theory of consolidation. It is obtained from laboratory consolidation tests and is used to predict the time-settlement behavior of field loading of fine-grained soil.

Cohesion   There are two types of cohesion: (1) cohesion in terms of total stress and (2) cohesion in terms of effective stress. For total cohesion (c), the soil particles are predominately held together by capillary tension. For effective stress cohesion (c'), there must be actual bonding or attraction forces between the soil particles.

Colloidal Soil Particles   Generally refers to clay size particles (finer than 0.002 mm) where the surface activity of the particle has an appreciable influence of the properties of the soil.

Compaction (Laboratory)

   Compaction Curve   A curve showing the relationship between the dry density and the water content of a soil for a given compaction energy.

   Compaction Test   A laboratory compaction procedure whereby a soil at a known water content is compacted into a mold of specific dimensions. The procedure is repeated for various water contents to establish the compaction curve. The most common testing procedures (compaction energy, number of soil layers in the mold, etc.) are the Modified Proctor (ASTM D 1557) or Standard Proctor (ASTM 698).

Compression Index   For a consolidation test, the slope of the linear portion of the vertical pressure versus void ratio curve on a semilog plot.

Consistency of Clay   Generally refers to the condition, in terms of firmness, of a clay. For example, a clay can have a consistency that varies from "very soft" up to "hard."

Consolidation Test   A laboratory test used to study the consolidation behavior of clay. The specimen is laterally confined in a ring and is compressed between porous plates (oedometer apparatus).

Contraction (during Shear)   During the shearing of soil, the tendency of loose soil to decrease in volume (or contract).

Density   Mass per unit volume. In the International System of Units (SI), typical units for the density of soil are Mg/m^3.

Deviator Stress   Difference between the major and minor principal stress in a triaxial test.

Dilation (during Shear)   During the shearing of soil, the tendency of dense soil to increase in volume (or dilate).

Double Layer   A grossly simplified interpretation of the positively charged water layer, together with the negatively charged surface of the particle itself. Three reasons for the attraction of water to the clay particle are (1) dipolar structure of water molecule which causes it to be electrostatically attracted to the surface of the clay particle, (2) hydrogen bonding which causes the water molecule to be held to the clay particle, and (3) the clay particles attract cations which contribute to the attraction of water by the hydration process.

Fabric (of Soil)   Definitions vary, but in general the fabric of soil often refers only to the geometric arrangement of the soil particles. In contrast, the soil structure refers to both the geometric arrangement of soil particles and the interparticle forces which may act between them.

Fine-Grained Soil According to the Unified Soil Classification System, a fine-grained soil contains more than 50 percent (by dry mass) of particles finer than the no. 200 sieve.

Flocculation When in suspension in water, the process of fines attracting each other to form a larger particle. In the hydrometer test, a dispersing agent is added to prevent flocculation of fines.

Friction Angle In terms of effective shear stress, the soil friction is usually considered to be due to the interlocking of the soil or rock grains and the resistance to sliding between the grains. A relative measure of a soil's frictional shear strength is the friction angle.

Gravel Size Fragments Rock fragments and soil particles that will pass the 3-in. (76-mm) sieve and be retained on a no. 4 (4.75-μm) U.S. standard sieve.

Hydraulic Conductivity (Coefficient of Permeability) For laminar flow of water in soil, both terms are synonymous and indicate a measure of the soil's ability to allow water to flow through its soil pores. The hydraulic conductivity is often measured in a constant-head or falling-head permeameter.

Laboratory Maximum Dry Density The peak point of the compaction curve (*see* Compaction).

Moisture Content (Water Content) Moisture content and water content are synonymous. The definition of moisture content is the ratio of the mass of water in the soil divided by the dry mass of the soil, usually expressed as a percentage.

Optimum Moisture Content The moisture content, determined from a laboratory compaction test, at which the maximum dry density of a soil is obtained by using a specific compaction energy.

Overconsolidation Ratio (OCR) The ratio of the preconsolidation vertical effective stress to the current vertical effective stress.

Peak Shear Strength The maximum shear strength along a shear failure surface.

Permeability The ability of water (or other fluid) to flow through a soil by traveling through the void spaces. A high permeability indicates that flow occurs rapidly, and vice versa. A measure of the soil's permeability is the hydraulic conductivity, also known as the *coefficient of permeability*.

Plasticity Term applied to silt and clay, to indicate the soil's ability to be rolled and molded without breaking apart. A measure of a soil's plasticity is the plasticity index.

Plasticity Index The plasticity index is defined as the liquid limit minus the plastic limit, often expressed as a whole number (*see* Atterberg Limits).

Sand Equivalent (SE) A measure of the amount of silt or clay contamination in fine aggregate as determined by ASTM D 2419 test procedures.

Sand-Size Particles Soil particles that will pass the no. 4 (4.75-mm) sieve and be retained on the no. 200 (0.075-mm) U.S. standard sieve.

Shear Strength   The maximum shear stress that a soil or rock can sustain. Shear strength of soil is based on total stresses (i.e., undrained shear strength) or effective stresses (i.e., effective shear strength).

Shear Strength in Terms of Total Stress   Shear strength of soil based on total stresses. The undrained shear strength of soil could be expressed in terms of the undrained shear strength s_u, or by using the failure envelope that is defined by total cohesion c and total friction angle ϕ.

Effective Shear Strength   Shear strength of soil based on effective stresses. The effective shear strength of soil could be expressed in terms of the failure envelope that is defined by effective cohesion c' and effective friction angle ϕ'.

Shear Strength Tests (Laboratory)   There are many types of shear strength tests that can be performed in the laboratory. The objective is to obtain the shear strength of the soil. Laboratory tests can generally be divided into two categories:

Shear Strength Tests Based on Total Stress   The purpose of these laboratory tests is to obtain the undrained shear strength of the soil or the failure envelope in terms of total stresses. An example is the unconfined compression test, which is also known as an *unconsolidated-undrained test.*

Shear Strength Tests Based on Effective Stress   The purpose of these laboratory tests is to obtain the effective shear strength of the soil based on the failure envelope in terms of effective stress. An example is a direct shear test where the saturated, submerged, and consolidated soil specimen is sheared slowly enough that excess pore water pressures do not develop (this test is known as a *consolidated-drained test*).

Sieve   Laboratory equipment consisting of a pan with a screen at the bottom. U.S. standard sieves are used to separate particles of a soil sample into their various sizes.

Silt-Size Particles   That portion of a soil that is finer than the no. 200 sieve (0.075 mm) and coarser than 0.002 mm. Silt and clay-size particles are considered to be "fines."

Soil Structure   Definitions vary, but in general, the soil structure refers to both the geometric arrangement of the soil particles and the interparticle forces which may act between them. Common soil structures are as follows:

Cluster Structure   Soil grains that consist of densely packed silt or clay-size particles.

Dispersed Structure   The clay-size particles are oriented parallel to each other.

Flocculated (or Cardhouse) Structure   The clay-size particles are oriented in edge-to-face arrangements.

Honeycomb Structure   Loosely arranged bundles of soil particles, having a structure that resembles a honeycomb.

Single-Grained Structure   An arrangement composed of individual soil particles. This is a common structure of sands.

Skeleton Structure An arrangement where coarser soil grains form a skeleton with the void spaces partly filled by a relatively loose arrangement of soil fines.

Specific Gravity The specific gravity of soil or oversize particles can be determined in the laboratory. Specific gravity is generally defined as the ratio of the density of the soil particles divided by the density of water.

Tensile Test For a geosynthetic, a laboratory test in which the geosynthetic is stretched in one direction to determine the force-elongation characteristics, breaking force, and the breaking elongation.

Texture (of Soil) The term texture refers to the degree of fineness of the soil, such as smooth, gritty, or sharp, when the soil is rubbed between the fingers.

Thixotropy The property of a remolded clay that enables it to stiffen (gain shear strength) in a relatively short time.

Triaxial Test A laboratory test in which a cylindrical specimen of soil or rock encased in an impervious membrane is subjected to a confining pressure and then loaded axially to failure.

Unconfined Compressive Strength The vertical stress which causes the shear failure of a cylindrical specimen of a plastic soil or rock in a simple compression test. For the simple compression test, the undrained shear strength s_u of the plastic soil is defined as one-half the unconfined compressive strength.

Unit Weight Unit weight is defined as weight per unit volume. In the International System of Units (SI), unit weight has units of kN/m^3 (kilonewtons per cubic meter). In the United States Customary System, unit weight has units of pcf (pounds force per cubic foot).

Water Content (Moisture Content) *See* Moisture Content.

Zero Air Voids Curve On the laboratory compaction curve, the zero air voids curve is often included. It is the relationship between water content and dry density for a condition of saturation ($S = 100\%$) for a specified specific gravity.

Glossary 3

Terminology for Engineering Analysis and Computations

Adhesion Shearing resistance between two different materials. For example, for piles driven into clay deposits, there is adhesion between the surface of the pile and the surrounding clay.

Allowable Bearing Pressure Allowable bearing pressure is the maximum pressure that can be imposed by a foundation onto soil or rock supporting the foundation. It is derived from experience and general usage, and provides an adequate factor of safety against shear failure and excessive settlement.

Anisotropic Soil A soil mass having different properties in different directions at any given point, referring primarily to stress-strain or permeability characteristics.

Arching The transfer of stress from an unconfined area to a less-yielding or restrained structure. Arching is important in the design of pile or pier walls that have open gaps between the members.

Bearing Capacity

    Allowable Bearing Capacity The maximum allowable bearing pressure for the design of foundations.

    Ultimate Bearing Capacity The bearing pressure that causes failure of the soil or rock supporting the foundation.

Bearing Capacity Failure A foundation failure that occurs when the shear stresses in the adjacent soil exceed the shear strength.

Bell The enlarged portion of the bottom of a drilled shaft foundation. A bell is used to increase the end bearing resistance. Not all drilled shafts have bells.

Collapsible Soil Collapsible soil can be broadly classified as soil that is susceptible to a large and sudden reduction in volume upon wetting. Collapsible soil usually has a low dry density and low moisture content. Such soil can withstand a large applied vertical stress with a small compression, but then experience much larger settlements after wetting, with no increase in vertical pressure. Collapsible soil can include fill compacted dry of optimum and natural collapsible soil, such as alluvium, colluvium, or loess.

Collapsible Formations Examples of collapsible formations include limestone formations and deep-mined coal beds. Limestone can form underground caves and caverns which can gradually enlarge resulting in a collapse of the ground surface and the formation of a sinkhole. Sites that are underlaid by coal or salt mines could also experience ground surface settlement when the underground mine collapses.

Compressibility A decrease in volume that occurs in the soil mass when it is subjected to an increase in loading. Some highly compressible soils are loose sands, organic clays, sensitive clays, highly plastic and soft clays, uncompacted fills, municipal landfills, and permafrost soils.

Consolidation The consolidation of a saturated clay deposit is generally divided into three separate categories:

Initial or Immediate Settlement The initial settlement of the structure caused by undrained shear deformations, or in some cases contained plastic flow, due to two- or three-dimensional loading.

Primary Consolidation The compression of clays under load that occurs as excess pore water pressures slowly dissipate with time.

Secondary Compression The final component of settlement, which is that part of the settlement that occurs after essentially all of the excess pore water pressures have dissipated.

Creep An imperceptibly slow and more or less continuous movement of slope-forming soil or rock debris.

Critical Height or Critical Slope Critical height refers to the maximum height at which a vertical excavation or slope will stand unsupported. Critical slope refers to the maximum angle at which a sloped bank of soil or rock of given height will stand unsupported.

Crown Generally, the highest point. For tunnels, the crown is the arched roof. For landslides, the crown is the area above the main scarp of the landslide.

Dead Load Structural loads due to the weight of beams, columns, floors, roofs and other fixed members. Does not include nonstructural items such as furniture, snow, occupants, and inventory.

Debris Flow An initial shear failure of a soil mass which then transforms itself into a fluid mass which can move rapidly over the ground surface.

Depth of Seasonal Moisture Change Also known as the *active zone,* the layer of expansive soil subjected to shrinkage during the dry season and swelling during the wet season. This zone extends from ground surface to the depth of significant moisture fluctuation.

Desiccation The process of shrinkage of clays. The process involves a reduction in volume of the grain skeleton and subsequent cracking of the clay caused by the development of capillary stresses in the pore water as the soil dries.

Design Load All forces and moments that are used to proportion a foundation. The design load includes the dead weight of a structure and, in some cases, can include live loads. Considerable judgment and experience are required to determine the design load that is to be used to proportion a foundation.

Downdrag Force induced on deep foundation resulting from downward movement of adjacent soil relative to foundation element. Also referred to as *negative skin friction.*

Earth Pressure Usually used in reference to the lateral pressure imposed by a soil mass against an earth-supporting structure such as a retaining wall or basement wall.

Active Earth Pressure k_a Horizontal pressure for a condition where the retaining wall has yielded sufficiently to allow the backfill to mobilize its shear strength.

At-Rest Earth Pressure k_o Horizontal pressure for a condition where the retaining wall has not yielded or compressed into the soil. This would also be applicable to a soil mass in its natural state.

Passive Earth Pressure k_p Horizontal pressure for a condition such as that resulting from a retaining wall footing that has moved into and compressed the soil sufficiently to develop its maximum lateral resistance.

Effective Stress The effective stress is defined as the total stress minus the pore water pressure.

Equipotential Line A line connecting points of equal total head.

Equivalent Fluid Pressure Horizontal pressures of soil, or soil and water in combination, which increase linearly with depth and are equivalent to those that would be produced by a soil of a given density. Equivalent fluid pressure is often used in the design of retaining walls.

Exit Gradient The hydraulic gradient near the toe of a dam or the bottom of an excavation through which groundwater seepage is exiting the ground surface.

Finite Element A soil and structure profile subdivided into regular geometrical shapes for the purpose of numerical stress analysis.

Flow Line The path of travel traced by moving groundwater as it flows through a soil mass.

Flow Net A graphical representation used to study the flow of groundwater through a soil. A flow net is composed of flow lines and equipotential lines.

Head From Bernoulli's energy equation, the total head is defined as the sum of the velocity head, pressure head, and elevation head. Head has units of length. For seepage problems in soil, the velocity head is usually small enough to be neglected, and thus, for laminar flow in soil, the total head h is equal to the sum of the pressure head h_p and elevation head h_z.

Heave The upward movement of foundations or other structures caused by frost heave or expansive soil and rock. Frost heave refers to the development of ice layers or lenses within the soil that cause the ground surface to heave upward. Heave due to expansive soil and rock is caused by an increase in water content of clays or rocks, such as shale or slate.

Homogeneous Soil Soil that exhibits essentially the same physical properties at every point throughout the soil mass.

Hydraulic Gradient Difference in total head at two points divided by the distance between them. Hydraulic gradient is used in seepage analyses.

Isotropic Soil A soil mass having essentially the same properties in all directions at any given point, referring primarily to stress-strain or permeability characteristics.

Laminar Flow Groundwater seepage in which the total head loss is proportional to the velocity.

Liquefaction The sudden, large decrease of shear strength of a cohesionless soil caused by collapse of the soil structure, produced by shock- or earthquake-induced shear strains, associated with a sudden but temporary increase of pore water pressures. Liquefaction causes the cohesionless soil to behave as a fluid.

Live Load Structural loads due to nonstructural members, such as furniture, occupants, inventory, and snow.

Mohr Circle A graphical representation of the stresses acting on the various planes at a given point in the soil.

Normally Consolidated The condition that exists if a soil deposit has never been subjected to an effective stress greater than the existing overburden pressure and if the deposit is completely consolidated under the existing overburden pressure.

Overconsolidated The condition that exists if a soil deposit has been subjected to an effective stress greater than the existing overburden pressure.

Piping The movement of soil particles as a result of unbalanced seepage forces produced by percolating water, leading to the development of ground surface boils or underground erosion voids and channels.

Plastic Equilibrium The state of stress of a soil mass that has been loaded and deformed to such an extent that its ultimate shearing resistance is mobilized at one or more points.

Pore Water Pressure The water pressure that exists in the soil void spaces.

Excess Pore Water Pressure The increment of pore water pressures greater than hydrostatic values, produced by consolidation stress in compressible materials or by shear strain.

Hydrostatic Pore Water Pressure Pore water pressure or groundwater pressures exerted under conditions of no flow where the magnitude of pore pressures increases linearly with depth below the ground surface.

Porosity The ratio, usually expressed as a percentage, of the volume of voids to the total volume of the soil or rock.

Preconsolidation Pressure The greatest vertical effective stress to which a soil, such as a clay layer, has been subjected.

Pressure (Stress) The load divided by the area over which it acts.

Principal Planes Each of three mutually perpendicular planes through a point in the soil mass on which the shearing stress is zero. For soil mechanics, compressive stresses are positive.

Major Principal Plane The plane normal to the direction of the major principal stress (highest stress in the soil).

Intermediate Principal Plane The plane normal to the direction of the intermediate principal stress.

Minor Principal Plane   The plane normal to the direction of the minor principal stress (lowest stress in the soil).

Principal Stresses   The stresses that occur on the principal planes. *See* Mohr Circle.

Progressive Failure   Formation and development of localized stresses which lead to fracturing of the soil, which spreads and eventually forms a continuous rupture surface and a failure condition. Stiff fissured clay slopes are especially susceptible to progressive failure.

Quick Clay   A clay that has a sensitivity greater than 16. Upon remolding, such clays can exhibit a fluid (or quick) condition.

Quick Condition (Quicksand)   A condition in which groundwater is flowing upward with a sufficient hydraulic gradient to produce a zero effective stress condition in the sand deposit.

Relative Density   Term applied to a sand deposit to indicate its relative density state, defined as the ratio of (1) the difference between the void ratio in the loosest state and *in situ* void ratio to (2) the difference between the void ratios in the loosest and in the densest states.

Saturation (Degree of)   The degree of saturation is calculated as the volume of water in the void space divided by the total volume of voids. It is usually expressed as a percentage. A completely dry soil has a degree of saturation of 0 percent and a saturated soil has a degree of saturation of 100 percent.

Seepage   The infiltration or percolation of water through soil and rock.

Seepage Analysis   An analysis to determine the quantity of groundwater flowing through a soil deposit. For example, by using a flow net, the quantity of groundwater flowing through or underneath a earth dam can be determined.

Seepage Force   The frictional drag of water flowing through the soil voids.

Seepage Velocity   The rate of discharge of seepage water through soil per unit area of void space perpendicular to the direction of flow.

Sensitivity   The ratio of the undrained shear strength of the undisturbed clay to the remolded shear strength of the same clay.

Settlement   The permanent downward vertical movement experienced by structures as the underlying soil consolidates, compresses, or collapses due to the structural load or secondary influences.

Differential Settlement   The difference in settlement between two foundation elements or between two points on a single foundation.

Total Settlement   The absolute vertical movement of the foundation.

Shear Failure   A failure in a soil or rock mass caused by shearing stain along a slip surface.

General Shear Failure   Failure in which the shear strength of the soil or rock is mobilized along the entire slip surface.

Local Shear Failure   Failure in which the shear strength of the soil or rock is mobilized only locally along the slip surface. Local shear failure can refer to punching type bearing capacity failures.

Progressive Shear Failure *See* Progressive Failure.

Shear Plane (Slip Surface) A plane along which failure of soil or rock occurs by shearing.

Shear Stress Stress that acts parallel to the surface element.

Slope Stability Analyses

Gross Slope Stability The stability of slope material below a plane approximately 0.9 to 1.2 m (3 to 4 ft) deep measured from and perpendicular to the slope face.

Surficial Slope Stability The stability of the outer 0.9 to 1.2 m (3 to 4 ft) of slope material measured from and perpendicular to the slope face. See Sec. 9.4 for typical repair of surficial failures.

Strain The change in shape of soil when it is acted upon by stress.

Normal Stain A measure of compressive or tensile deformations, defined as the change in length divided by the initial length. In geotechnical engineering, strain is positive when it results in compression of the soil.

Shear Strain A measure of the shear deformation of soil.

Subsidence Settlement of the ground surface over a very large area, such as that caused by the extraction of oil from the ground or the pumping of groundwater from wells.

Swell Increase in soil volume, typically referring to volumetric expansion of clay due to an increase in water content.

Time Factor *T* A dimensionless factor, used in the theory of consolidation or swelling of clay.

Total Stress The effective stress plus the pore water pressure. The vertical total stress can be calculated by multiplying the total unit weight of the soil by the depth below ground surface.

Underconsolidation The condition that exists if a soil deposit is not fully consolidated under the existing overburden pressure and excess pore water pressures exist within the material. Underconsolidation can occur in areas where a clay is being deposited very rapidly and not enough time has elapsed for the soil to consolidate under the overburden pressure.

Void Ratio The volume of voids divided by the volume of soil particles.

Glossary 4

Construction and Grading Terminology

Aggregate A granular material used for a pavement base, wall backfill, etc.

    Coarse Aggregate Gravel or crushed rock that is retained on the no. 4 sieve (4.75 mm).

    Fine Aggregate Often refers to sand (passes the no. 4 sieve and is retained on the no. 200 U.S. standard sieve).

    Open-Graded Aggregate Generally refers to a gravel that does not contain any soil particles finer than the no. 4 sieve.

Apparent Opening Size For a geotextile, a property which indicates approximately the largest particle that would effectively pass through the geotextile.

Approval A written engineering or geological opinion by the responsible engineer, geologist of record, or responsible principal of the engineering company concerning the process and completion of the work, unless it specifically refers to the building official.

Approved Plans The current grading plans which bear the stamp of approval of the building official.

Approved Testing Agency A facility whose testing operations are controlled and monitored by a registered civil engineer and which is equipped to perform and certify the tests as required by the local building code or building official.

As-Graded (As-Built) The surface conditions at the completion of grading.

Asphalt A dark brown to black cementitious material whose main ingredient is bitumen (high-molecular-weight hydrocarbons) that occurs in nature or is obtained from petroleum processing.

Asphalt Concrete (AC) A mixture of asphalt and aggregate that is compacted into a dense pavement surface. Asphalt concrete is often prepared in a batch plant.

Backdrain Generally a pipe and gravel or similar drainage system placed behind earth-retaining structures such as buttresses, stabilization fills, and retaining walls.

Backfill Soil material placed behind or on top of an area that has been excavated. For example, backfill is placed behind retaining walls and in utility trench excavations.

Base Course or Base A layer of specified or selected material of planned thickness constructed on the subgrade or subbase for the purpose of providing

support to the overlying concrete or asphalt concrete surface of roads and airfields.

Bench   A relatively level step excavated into earth material on which fill is to
be placed.

Berm   A raised bank or path of soil. For example, a berm is often constructed at
the top of slopes to prevent water from flowing over the top of the slope.

Borrow   Earth material acquired from an off-site location for use in grading on
a site.

Brooming   The crushing or separation of wood fibers at the butt (top of the
pile) of a timber pile while it is being driven.

Building Official   The city engineer, director of the local building department,
or duly delegated representative.

Bulking   The increase in volume of soil or rock caused by its excavation. For
example, rock or dense soil will increase in volume upon excavation or by being
dumped into a truck for transportation.

Buttress Fill   A fill mass, the configuration of which is designed by engineering
calculations to stabilize a slope exhibiting adverse geologic features. A buttress
is generally specified by minimum key width and depth and by maximum backcut angle. A buttress normally contains a backdrainage system.

Caisson   Sometimes large-diameter piers are referred to as caissons. Another
definition is a large structural chamber utilized to keep soil and water from
entering into a deep excavation or construction area. Caissons may be installed
by being sunk in place or by excavating the bottom of the unit as it slowly sinks
to the desired depth.

Cat   Slang for Caterpillar grading or construction equipment.

Clearing, Brushing, and Grubbing   The removal of vegetation (grass, brush,
trees, and similar plant types) by mechanical means.

Clogging   For a geotextile, a decrease in permeability due to soil particles that
have either logged in the geotextile openings or have built up a restrictive layer
on the surface of the geotextile.

Compaction   The densification of a fill by mechanical means.

Compaction Equipment   Compaction equipment can be grouped generally into
five different types or classifications: sheepsfoot, vibratory, pneumatic, high-
speed tamping foot, and chopper wheels (for municipal landfill). Combinations
of these types are also available.

Compaction Production   Compaction production is expressed in compacted
cubic meters (m^3) or compacted cubic yards (yd^3) per hour.

Concrete   A mixture of aggregates (sand and gravel) and paste (Portland
cement and water). The paste binds the aggregates together into a rocklike
mass as the paste hardens because of the chemical reactions between the
cement and the water.

Contractor   A person or company under contract or otherwise retained by the client to perform demolition, grading, and other site improvements.

Cut/Fill Transition   The location in a building pad where on one side the pad has been cut down, exposing natural or rock material, while on the other side, fill has been placed.

Dam   A structure built to impound water or other fluid products such as tailing waste, wastewater effluent, etc.

   Homogeneous Earth Dam   An earth dam whose embankment is formed of one soil type without a systematic zoning of fill materials.

   Zoned Earth Dam   An earth dam embankment zoned by the systematic distribution of soil types according to their strength and permeability characteristics, usually with a central impervious core and shells of coarser materials.

Debris   All products of clearing, grubbing, demolition, or contaminated soil material that are unsuitable for reuse as compacted fill and/or any other material so designated by the geotechnical engineer or building official.

Dewatering   The process used to remove water from a construction site, such as pumping from wells in order to lower the groundwater table during a foundation excavation.

Dozer   Slang for bulldozer construction equipment.

Drainage   The removal of surface water from the site.

Drawdown   The lowering of the groundwater table that occurs in the vicinity of a well that is in the process of being pumped.

Earth Material   Any rock, natural soil, or fill, or any combination thereof.

Electroosmosis   A method of dewatering, applicable for silts and clays, in which an electric field is established in the soil mass to cause the movement by electroosmotic forces of pore water to wellpoint cathodes.

Erosion Control Devices (Temporary)   Devices which are removable and can rarely be salvaged for subsequent reuse. In most cases they will last no longer than one rainy season. They include sandbags, gravel bags, plastic sheeting (visqueen), silt fencing, straw bales, and similar items.

Erosion Control System   A combination of desilting facilities and erosion protection, including effective planting to protect adjacent private property, watercourses, public facilities, and receiving waters from any abnormal deposition of sediment or dust.

Excavation   The mechanical removal of earth material.

Fill   A deposit of earth material placed by artificial means. An *engineered* (or *structural*) *fill* refers to a fill in which the geotechnical engineer has, during grading, made sufficient tests to enable the conclusion that the fill has been placed in substantial compliance with the recommendations of the geotechnical engineer and the governing agency requirements.

   Hydraulic Fill   A fill placed by transporting soils through a pipe using large quantities of water. These fills are generally loose because they have little or no mechanical compaction during construction.

Footing   A structural member typically installed at a shallow depth that is used to transmit structural loads to the soil or rock strata. Common types of footings include combined footings, spread (or pad) footings, and strip (or wall) footings.

Forms   Usually made of wood, forms are used during the placement of concrete. Forms confine and support the fluid concrete as it hardens.

Foundation   That part of the structure that supports the weight of the structure and transmits the load to underlying soil or rock.

   Deep Foundation   A foundation that derives its support by transferring loads to soil or rock at some depth below the structure.

   Shallow Foundation   A foundation that derives its support by transferring load directly to soil or rock at a shallow depth.

Freeze   Also known as *setup,* an increase in the load capacity of a pile after it has been driven. Freeze is caused primarily by the dissipation of excess pore water pressures.

Geosynthetic   A planar product manufactured from polymeric material and typically placed in soil to form an integral part of a drainage, reinforcement, or stabilization system.

Geotextile   A permeable geosynthetic composed solely of textiles.

Grade   The vertical location of the ground surface.

   Existing Grade   The ground surface prior to grading.

   Finished Grade   The final grade of the site which conforms to the approved plan.

   Lowest Adjacent Grade   Adjacent the structure, the lowest point of elevation of the finished surface of the ground, paving, or sidewalk.

   Natural Grade   The ground surface unaltered by artificial means.

   Rough Grade   The stage at which the grade approximately conforms to the approved plan.

Grading   Any operation consisting of excavation, filling, or combination thereof.

Grading Contractor   A contractor licensed and regulated who specializes in grading work or is otherwise licensed to do grading work.

Grading Permit   An official document or certificate issued by the building official, authorizing grading activity as specified by approved plans and specifications.

Grouting   The process of injecting grout into soil or rock formations to change their physical characteristics. Common examples include grouting to decrease the permeability of a soil or rock strata and compaction grouting to densify loose soil or fill.

Hillside Site   A site which entails cut and/or fill grading of a slope which may be adversely affected by drainage and/or stability conditions within or outside the site, or which may cause an adverse affect on adjacent property.

Jetting The use of a water jet to facilitate the installation of a pile. It can also refer to the fluid placement of soil, such as jetting in the soil for a utility trench.

Key A designed compacted fill placed in a trench excavated in earth material beneath the toe of a proposed fill slope.

Keyway An excavated trench into competent earth material beneath the toe of a proposed fill slope.

Lift During compaction operations, a lift is a layer of soil that is dumped by the construction equipment and then subsequently compacted as structural fill.

Necking A reduction in cross-sectional area of a drilled shaft as a result of the inward movement of the adjacent soils.

Owner Any person, agency, firm, or corporation having a legal or equitable interest in a given real property.

Permanent Erosion Control Devices Improvements which remain throughout the life of the development. They include terrace drains, downdrains, slope landscaping, channels, storm drains, etc.

Permit An official document or certificate issued by the building official authorizing performance of a specified activity.

Pier A deep foundation system, similar to a cast-in-place pile, that consists of columnlike reinforced-concrete members. Piers are often of large enough diameter to enable downhole inspection. Piers are also commonly referred to as *drilled shafts, bored piles,* or *drilled caissons.*

Pile A deep foundation system, consisting of relatively long, slender, columnlike members that are often driven into the ground.

  Batter Pile A pile driven in at an angle inclined to the vertical to provide higher resistance to lateral loads.

  Combination End-Bearing and Friction Pile A pile that derives its capacity from combined end-bearing resistance developed at the pile tip and frictional and/or adhesion resistance on the pile perimeter.

  End-Bearing Pile A pile whose support capacity is derived principally from the resistance of the foundation material on which the pile tip rests.

  Friction Pile A pile whose support capacity is derived principally from the resistance of the soil friction and/or adhesion mobilized along the side of the embedded pile.

Pozzolan For concrete mix design, a siliceous or siliceous and aluminous material which will chemically react with calcium hydroxide within the cement paste to form compounds having cementitious properties.

Precise Grading Permit A permit that is issued on the basis of approved plans which show the precise structure location, finish elevations, and all on-site improvements.

Relative Compaction The degree of compaction (expressed as a percentage) defined as the field dry density divided by the laboratory maximum dry density.

Rippability A characteristic of rock or dense and rocky soil by which it can be excavated without blasting. Excavation, or ripping, is accomplished by using equipment such as a Caterpillar ripper, ripper-scarifier, tractor-ripper, or impact ripper. Ripper performance has been correlated with the seismic wave velocity of the soil or rock (see *Caterpillar Performance Handbook*, 1997).

Riprap   Rocks that are generally less than 1800 kg (2 tons) in mass that are placed on the ground surface, on slopes or at the toe of slopes, or on top of structures to prevent erosion by wave action or strong currents.

Running Soil or Running Ground   In tunneling or trench excavations, a granular material that tends to flow or "run" into the excavation.

Sand Boil   The ejection of sand at ground surface, usually forming a cone shape, caused by underground piping.

Shear Key   Similar to a buttress, but generally constructed by excavating a slot within a natural slope in order to stabilize the upper portion of the slope without grading encroachment into the lower portion of the slope. A shear key is also often used to increase the factor of safety of an ancient landslide.

Shrinkage Factor   When the loose material is worked into a compacted state, the shrinkage factor (SF) is the ratio of the volume of compacted material to the volume of borrow material.

Shotcrete   Mortar or concrete pumped through a hose and projected at high velocity onto a surface. Shotcrete can be applied by a "wet" or "dry" mix method.

Site   The particular lot or parcel of land where grading or other development is performed.

Slope   An inclined ground surface. For graded slopes, the steepness is generally specified as a ratio of horizontal:vertical (e.g., 2:1 slope). Common types of slopes include natural (unaltered) slopes, cut slopes, false slopes (temporary slopes generated during fill compaction operations), and fill slopes.

Slough   Loose, noncompacted fill material generated during grading operations. Slough can also refer to a shallow slope failure, such as sloughing of the slope face.

Slump   In the placement of concrete, the slump is a measure of consistency of freshly mixed concrete as measured by the slump test. In geotechnical engineering, a slump could also refer to a slope failure.

Slurry Seal   In the construction of asphalt pavements, a slurry seal is a fluid mixture of bituminous emulsion, fine aggregate, mineral filler, and water. A slurry seal is applied to the top surface of an asphalt pavement in order to seal its surface and prolong its wearing life.

Soil Stabilization   The treatment of soil to improve its properties. There are many methods of soil stabilization such as adding gravel, cement, or lime to the soil. The soil could also be stabilized by using geotextiles, by drainage, or by compaction.

Specification   A precise statement in the form of specific requirements. The requirements could be applicable to a material, product, system, or engineering service.

Stabilization Fill Similar to a buttress fill, the configuration of which is typically related to slope height and is specified by the standards of practice for enhancing the stability of locally adverse conditions. A stabilization fill is normally specified by minimum key width and depth and by maximum backcut angle. A stabilization fill usually has a backdrainage system.

Staking During grading, staking is the process in which a land surveyor places wood stakes that indicate the elevation of existing ground surface and the final proposed elevation per the grading plans.

Structure A structure is defined as that which is built or constructed, an edifice or building of any kind, or any piece of work artificially built up or composed of parts joined together in some definite manner.

Subdrain A pipe-and-gravel or similar drainage system placed in the alignment of canyons or former drainage channels. After placement of the subdrain, structural fill is placed on top of the subdrain.

Subgrade For roads and airfields, the subgrade is defined as the underlying soil or rock that supports the pavement section (subbase, base, and wearing surface). The subgrade is sometimes referred to as the *basement soil* or *foundation soil*.

Substructure and Superstructure The substructure is the foundation and the superstructure is the portion of the structure located above the foundation (includes beams, columns, floors, and other structural and architectural members).

Sulfate (SO$_4$) A chemical compound occurring in some soils which, above certain levels of concentration, has a corrosive effect on ordinary Portland cement concrete and some metals.

Sump A small pit excavated in the ground or through the basement floor to serve as a collection basin for surface runoff or groundwater. A sump pump is used to periodically drain the pit when it fills with water.

Tack Coat In the construction of asphalt pavements, the tack coat is a bituminous material that is applied to an existing surface to provide a bond between different layers of the asphalt concrete.

Tailings In terms of grading, tailings is nonengineered fill which accumulates on or adjacent to equipment haul roads.

Terrace A relatively level step constructed in the face of a graded slope surface for drainage control and maintenance purposes.

Underpinning Piles or other types of foundations built to provide new support for an existing foundation. Underpinning is often used as a remedial measure.

Vibrodensification The densification or compaction of cohesionless soils by imparting vibrations into the soil mass so as to rearrange soil particles resulting in less voids in the overall mass.

Walls

Bearing Wall Any metal or wood stud walls that support more than 100 lb per linear foot of superimposed load. Any masonry or concrete wall that supports more than 200 lb per linear foot of superimposed load or is more than one story high (UBC, 1997).

Cutoff Wall   The construction of tight sheeting or a barrier of impervious material extending downward to an essentially impervious lower boundary to intercept and block the path of groundwater seepage. Cutoff walls are often used in dam construction.

Retaining Wall   A wall designed to resist the lateral displacement of soil or other materials.

Water-Cement Ratio   For concrete mix design, the ratio of the mass of water (exclusive of that part absorbed by the aggregates) to the mass of cement.

Well Point   During the pumping of groundwater, the well point is the perforated end section of a well pipe where the groundwater is drawn into the pipe.

Windrow   A string of large rock buried within engineered fill in accordance with guidelines set forth by the geotechnical engineer or governing agency requirements.

Workability of Concrete   The ability to manipulate a freshly mixed quantity of concrete with a minimum loss of homogeneity.

Index

ABOUT THE AUTHOR

Robert W. Day is a leading forensic engineer and the chief engineer at American Geotechnical in San Diego, California. The author of over 200 published technical papers, he serves on advisory committees for several professional associations, including ASCE, ASTM, and NCEES. He holds four college degrees: two from Villanova University, bachelor's and master's degrees majoring in structural engineering; and two from the Massachusetts Institute of Technology, master's and the *Civil Engineer* degree (highest degree) majoring in geotechnical engineering. He is also a registered civil engineer in several states and a registered geotechnical engineer in California.